Fields of Farmers:
Interning, Mentoring, Partnering, Germinating

Fields of Farmers:
Interning, Mentoring, Partnering, Germinating

by Joel Salatin

Polyface Inc.
Swoope, Virginia

Fields of Farmers: Interning, Mentoring, Partnering, Germinating, First Edition
Copyright © 2013 Joel Salatin

Editing by Lindsay Curren and book design by Jennifer Dehoff

About the Cover: Image by Rachel Salatin, with her explanation:
I imagine farms and homesteads of the past being a variety of different sizes, densely scattered across America and being tended by individuals and families just as dissimilar. The cover photo illustrates this concept of diversity and density through a thick grass field at Polyface Farm. Just like the assortment of grasses, farming allows for individuals from many professions, with unique ideals and interests to prosper. The cattle grazing in the background whimsically illustrates the romantic outcome of the hardworking farmer in the foreground.

Library of Congress Control Number: 2013912674

ISBN: 9780963810977

Other Books By Joel Salatin

Pastured Poultry Profit$:
Net $25,000 on 20 Acres in 6 Months

$alad Bar Beef

You Can Farm:
The Entrepreneur's Guide to Start and $ucceed in A Farming Enterprise

Family Friendly Farming:
A Multi-Generational Home-Based Business Testament

Holy Cows and Hog Heaven:
The Food Buyer's Guide to Farm Friendly Food

Everything I Want to Do Is Illegal:
War Stories from the Local Food Front

The Sheer Ecstasy of Being a Lunatic Farmer

Folks, This Ain't Normal:
A Farmer's Advice for Happier Hens, Healthier People, and a Better World

All Books Available From:

Polyface Gift Shop   1-540-885-3590
www.polyfacefarms.com

Chelsea Green Publishing   1-800-639-4099
www.chelseagreen.com

Acres USA Magazine   1-800-355-5313
www.acresusa.com

Amazon.com   www.amazon.com

Your local bookstore

TABLE OF CONTENTS

Acknowledgements

I've cried a few times writing this book. At times I've simply been overwhelmed with gratitude for my wife, Teresa, whose loyalty and encouragement have lifted me when I was down.

Fortunately, when one of us is down, the other is up. I can't think of a time when both of us were down. What a gal.

My mother, Lucille, known by everyone at Polyface affectionately as "Grandma" continues to encourage, lead and challenge. Although he passed away in 1988, my dad holds the genius visionary spot in my heart. Together he and mom created a nest for Teresa and I to grow. They germinated us in this field of farmers.

Our son Daniel, who now runs day-to-day operations on the farm, has made this book possible; along with his wife, Sheri. Their leadership and spirit literally buoy the farm's energy every day. Their three children and my grandchildren, Travis, Andrew, and Lauryn are the future. They keep me going.

Our daughter Rachel, as usual came through with a great cover for the book. Our immediate family presents a wonderful study in dynamics for both the opportunities and challenges we face as we seek abundance in our most intimate relationships.

All of our former interns, apprentices, and current farm subcontractors, many of whom you will meet in these pages, have enriched my life with experiences and wisdom that this book imparts.

Finally, our staff works tirelessly so I can write and think: Eric, Leanna, Brie, Wendy, Richard, Jackie and apprentices Noah, Jonathon, Ben, and Heather. Editing by Lindsay Curren and layout by Jennifer Dehoff created the final product you have in your hands.

Foreword

As I write this in mid-2013 there are 26 million unemployed young people in the rich countries of the world. In the USA youth unemployment ranges from 25 percent for young whites to 40 percent for young American blacks. In Spain, youth unemployment is a staggering 56 percent! However, there are three developed countries where youth unemployment is in the four to six percent range. These countries are Austria, Switzerland and Germany. What separates them from the rest of the developed world is their system of dual-education.

Young people in these countries decide at age 14 whether they want to continue with the formal education path that leads to a university degree or whether they want to choose a vocational path that combines paid intern work with learning and leads to a certified skilled trade. These vocational trades cover some 350 specialties and include sales, marketing, accounting, shipping (logistics) and agriculture. Currently, two-thirds of the young people in Germany choose the vocational education route.

The young people on the vocational track work three or four days a week for a firm or farm that pays them a sub-minimum intern's wage and teaches them the relevant skills needed in that industry. Chambers of Commerce and trade associations make sure that the work and teaching are matched and technologically up to date. After working for three years, the interns are certified and the best are usually offered a position by the firm they interned with.

While many American education specialists criticize having young people make a career decision at 14, the decision age was originally set at 10 in the late 1960s but was changed recently due to professional educators' pressure. However, some psychological studies have indicated that the career you wanted at 10 is the one most likely to make you happy and fulfilled in life. That many parents and teachers try to push a child into a career of *their* choosing rather than the child's apparently can haunt one for a lifetime. Also, starting to work at something early in life appears to be critical to one's lifetime earnings.

According to an American study quoted by the British financial magazine *The Economist*, beginning to work early in life is a key component in obtaining a higher than average lifetime wage. The study found that young people who never worked after school in High School faced a wage penalty of 20 percent for 20 years. High School dropouts with no work experience find getting an initial job a major challenge and many turn to a life of crime out of frustration.

The Economist said the problem of developed world, youth unemployment is rooted in a too high minimum wage and the disconnect between formal education and real-world job needs. For full employment to exist, the employee's pay must be tied to the value the employer receives from the worker's labor regardless of the employee's age.

As Joel Salatin repeatedly points out in this book, education is expensive. At one time, all employers took on the responsibility for training their employees but now expect someone else, somewhere else, to do this for them. Most employers far prefer to poach a proven worker from a competitor than to go to the expense of developing their own. This is becoming increasingly difficult. Currently, there is a major push by industry to open up immigration to work-proven foreigners due to employers' frustration with today's poorly trained worker pool.

And it is not just the employers who are unhappy with our education system. Currently half of American college graduates hold a job that does not require a college degree. Federal student loans have increased by 60 percent in the last five years while the average pay for college graduates has fallen by five percent in the same time period according to *The Wall Street Journal*. One in five American households are still paying on their student debt.

I am like Joel in that I loved my time in college but I must admit that all of the relevant career useful information I learned could have easily been taught in six weeks rather four years. What saved me was that I could take a full load of classes and still work four hours a day in my chosen field after class. When I graduated, I just went to eight hours a day and never had to look for a job. I have subsequently never had an employer ask if I had a college degree. They were only interested in where I had worked and what I had done. Going to an exclusive "name" university may help you to get a better paying first job but after that first job your pay is going to be based upon your work experience and the results you have produced.

As an employer I have learned how frustrating it is to be the first employer of a newly minted college grad. Their expectations about initial pay and responsibility are totally unrealistic. They just don't realize how little they know that is useful and this is not all their fault.

I was invited once to address a group of journalism students at our local university and explained that they should see the next ten years of their career as an internship. I said they should try to go to work for the smallest newspaper they could find because there they would be given the chance to do lots of different things and could explore different job options. I told them that after completing such an internship they would then, and only then, have the skills to start to make some serious money.

Needless to say, my talk went over like a lead balloon with the students but what really surprised me was how upset the professor was. He said I shouldn't say such things because the students might get discouraged and change their major. It was then that I realized where those misconceptions I was seeing in recent graduates were coming from. These young people were not being prepared for the realities of the real world. They were being fed a cake baked by Rosy Scenario to keep the enrollment numbers up.

The bottom line here is that the average college graduate does not make more than a high school graduate until she is 35 years old! That a college degree does not give you a pass on the need for an introductory internship in the real world is something young people must discover on their own today because the "educators" are not going to tell them.

Some educators criticize early vocational education because the trade young people learn may not be relevant in the future and this is true. Today's high tech is tomorrow's scrap heap. However, colleges and universities have a very hard time teaching cutting edge technology because of the constraints of the accreditation boards. By the time a PhD is available to teach a cutting edge technological subject it is already out of date. This is why industry must train their own. They should be willing to offer internships because that way they will get first shot at permanently hiring the really good ones.

This is why I think a college education should primarily teach skills that are needed and useful regardless of the career path chosen. Accounting, sales, marketing, reading and writing for comprehension, negotiation skills and the basics of consumer psychology are universal and will never go out of date. My father believed that agriculture was the one industry that would never go out of date because eating was the last thing people would give up. A study of businesses that have lasted for over 500 years in Europe found that almost all of them were based upon an agricultural product. While the specific technological tools change, the core interface of sun and soil doesn't.

Several European countries are studying adopting the dual-education system. This is being vociferously opposed by professional educators and union officials but they have no alternative to counter with. No doubt it is only a matter of time until this revolution in education arrives in mainstream North America.

A writer in *The Wall Street Journal* recently said he thought that American higher education is right where the Christian church was just before the Reformation. It has become too expensive, is self-serving and no longer delivers on its promises. What the details of this change will be are beyond me but I have no doubt that internships will be a big part of the new era of American education.

As usual, Joel Salatin is once again on the cutting edge of change. Whether you want to be an intern, or hire an intern, reading this book will be invaluable for you.

Allan Nation
Editor and Co-owner
The Stockman Grass Farmer

Introduction

bundance. What a wonderful word. To associate it with farming may seem inappropriate or naive. And yet that is exactly what I see every day on our farm. Stepping out into the early morning light, inhaling the fragrance of a billion red clover blossoms, feasting my eyes on slick, fat cows and succulent garden tomatoes, I'm reminded every morning that this ecological womb produces prodigious life.

Few people enjoy participating this viscerally in the happiness of as many beings as I do. When I move the cows to a new paddock, they kick up their heels and dance, cavorting sideways and demonstrating sheer joy. Moving the chickens to a new paddock excites a revelry ritual of chasing down bugs, finding new worms, and gorging on dew-speckled grass and clover. How many people get to make this many beings happy every day?

I can almost hear a garden symphony as I trellis tomatoes, thin the carrots, and hill the corn. Potato bugs will not rob me of a bountiful crop; I shake them into a can to release the potato plants and give the bugs to the chickens, who come running when I approach, knowing that my presence brings tasty morsels. It might be kitchen scraps or potato bugs, but it's all like homemade ice cream to the chickens.

As I go about chores, I greet my grandchildren taking care of their own farm enterprises, bringing more abundance to this landscape. The intern crew, busily caring for animals and plants; the farm staff discussing what to plant next or how to serve another restaurant--the bustling and happiness indicate an emotionally, economically, and ecologically enhancing place.

Knowing that this year we can build more ponds to hold more raindrops where they fall; that we can give bonuses to our team; that we can put a little more in savings--this is all abundance. Driven by sunlight, leveraged with ingenuity, developed in relationships, this abundance radiates satisfaction, blessing, and joy.

I find it difficult to relate to people who describe farms as places of drudgery, poverty, and burden. The miracle of life, whether it's a chick hatching, cow calving, or seed sprouting, draws us irresistibly to the wonder of our nest. That so many people think this visceral relationship with life's daily wonders is not attractive or appropriate for technologically advanced sophisticates indicates a profound hubris and lack of understanding. We are all utterly and completely dependent on soil, honey bees, raindrops, sunlight, fungi, and bacteria. Neither the greatest scientific discovery nor the highest gain on Wall Street compares to the importance of a functioning carbon cycle or dancing earthworms.

Farms can be profitable. They can be emotionally rewarding. They can be places of abundant living in all its aspects. They can be places where relationships thrive, where beauty abounds, and where healing occurs. The developed world has spent a couple of generations now marginalizing and pooh-poohing the idea of an abundant farm. Too many are not.

Many farmers don't think it's possible. Result? The average farmer is now 60 years old. Many young people, though, growing up in an era of environmental awareness and increasingly food savvy, crave a connection to this abundant womb . . . with their hands in the soil and heads in appropriate innovation. The earth is ready for these new stewards. Our food system needs them.

The imperative of our day is to create multi-generational land-based enterprises of abundance. In a time of increasing fear, perhaps nothing could be more needed or noble than stimulating abundance with a new generation of farmers. Turning deserts into verdant foodscapes can be done without depleting aquifers. Turning unprofitable, dead-end farms into vibrant food clusters can be done without chemical fertilizers, cheap fuel, and genetically modified organisms.

Indeed, our extractive production paradigm can be replaced with a regeneration model: building soil, strengthening nutrient density, hydrating a parched landscape, and germinating new farmers. If you own land and don't know what to do with it, this book is for you. If you want to farm, but don't know how to start, this book is for you. If you are an aging farmer struggling with a succession plan, this book is for you. And if you are a farmer's child trying to make a place for yourself on the family farm, this book is for you.

As I approach that 60-year point and contrast what I enjoy every day compared to the struggles I see in most other farmers my age, I realize how special our Polyface Farm really is. Although it's as wonderful as a fairy tale, it's not a fairly tale. It's the physical expression of the spirit and activities we've used for a couple of generations. We've run a formal apprentice program now for nearly 20 years. As I bare my heart and soul through this book, realize I'm sharing a lifetime of successes and failures. This is my best advice right now. I'm sure I'll learn something more tomorrow, but this is what I think today.

Although this book is extremely practical and full of how-to, the overriding theme is about creating abundance in our lives, on our farms, and for creation. May we enjoy an abundant future filled with *Fields of Farmers*.

Joel Salatin
Summer, 2013

What's the Big Deal?

Chapter 1

The Need

W ho will produce our food? Who will steward the land? We are living in a time of unsettling convergence. On the one hand, world population is at an historical high and needs food. We need farmers. But worldwide, and perhaps most acutely in the U.S. and Australia, the average farmer is now about 60 years old. According to agricultural statisticians, in the next couple of decades, nearly 50 percent of farmland will change hands.

Anyone who knows anything about farming today is aware of this accelerating phenomenon. Into whose hands will this farmland go? Perhaps one of the most interesting studies I've seen was Cornell University's study of abandoned farmland. Not farmland converted to strip malls, roads, or houses, but abandoned from active farming. It looked at a 15-year stretch from about 1995-2010 and found that in New York state alone some 3.1 million acres fit the category.

If you drive through upstate New York today, you'll pass mile upon mile of rich farmland now in the early successional stages of reverting to forest. Brambles, saplings, and decomposing barns fill a landscape once alive with dairies and farm-dependent communities. The people who live there often work elsewhere and generally eat the same food as their inner city cousins. Although these communities are imbedded in some of the richest land on the planet, for all practical purposes, they are more akin to inner city food deserts than the self-reliant abundance of their ancestors.

Everywhere I go, from California to Australia to New York to Mississippi, the abandoned farmland phenomenon is apparent. The issue of farmland abandonment now looms larger as a cultural conundrum than farmland development, which spawned the farmland preservation and conservation easement movements in the late 1980s.

We're in a race of time against this aging farm population, not just from a food production and land stewardship angle, but also from a functional society angle. The ramifications of large scale rural land abandonment has been best documented by the Center for Rural Affairs in Nebraska. Rather than repeat all the issues, I'd like to move directly to the compelling question, *Who* will touch this land?

Almost every day I receive a letter from a farm owner that goes something like this: "My wife and I are aging and our children don't want to farm this land that has been in our family for three generations. I'm 75 years old and simply can't keep it up. Our middle-aged children don't want us to sell the farm, but their careers keep them too busy to take an active part and help me out. I can't get up and down from the tractor like I used to. Getting in the corral with the cows is becoming unsafe for me and my wife. Can you send us a young person who will partner with us and care take this farm for our family?"

Doesn't that break your heart? Another frequent theme is: "My husband and I will inherit a farm but we have careers elsewhere and want to keep the farm in the family. We don't want to be the generation to lose the farm, which our family has owned for more than a century. Is there a way for a young person to actually make it on this farm? My parents did, but all I remember is hard work, dust, and obnoxious odors. I couldn't get away fast enough. Now I wish I had taken more interest, but my husband and I are in our early 50s and not in a position to go back to the farm. We desperately need a young partner to work alongside my parents before they pass away, to learn what my parents know, and to eventually manage this beautiful farm. Can you help us?"

And here is an increasingly common plea: "My husband and I made good money during the e-boom and always wanted a piece of land. Our financial picture enabled us to buy a farm but we're floundering a bit. We now realize it's not as easy as we thought, that it takes more skill and understanding than we thought. And we've kind of gotten used to traveling so we don't want to be tied down to the farm. Do you have an intern who could come and manage our farm? We have equipment, buildings, and equity; what we need is know-how and youthful energy. Can you help us?"

Each story has its own twist, but you get the picture. Sometimes the needs are more sophisticated, such as urban nonprofits looking for someone to manage an inner-city micro-farm. Energy costs and food quality concerns converge to stimulate interest in urban farming, which offers reduced food miles and re-localization. Nearly every time I speak in an urban setting, someone will come up to me with a story about a family farm in jeopardy and beg for one of our interns to come and help them out.

Although I haven't gotten many letters like this, I've even received letters from very elderly farmers asking me to find a young person that they can pass their farm to. One I'll never forget went, "I'm 85 years old. None of my three children has taken an interest in the farm, except to cash it out when I'm gone. I love this land. I love this farm. I've spent my life here on these 125 acres, helped my children through college, and thrived on the work and seasons of this spot. I think it's close to paradise. But if I give it to my children, they'll sell it for development. Can you find me a young person I can inherit it to?"

Folks, I'm not making any of this up. If I printed verbatim the distress signals that come across my desk every day, you couldn't read the words through the tears in your eyes. Few things capture the summation of our memories, our relationships, and the progress of our lives like the land we love. We remember the tree with the swing, and the day the tree blew down in a windstorm. Building a little dam with rocks in the creek. Lying on our back, with that favorite cousin, enjoying cozy conversation and watching a circling vulture overhead. These form life centerpieces.

These experiences and the emotions they engender span all kinds of farms. This isn't a time to judge about whether a farmer used DDT forty years ago. It's not the time for self-righteous finger pointing from an urban Lay-Z-Boy about the rightness or wrongness of feeding grain to cows. No, this is about real people who formed the backbone of rural America, pledged to feed a desperate post-WWII world, and contributed countless charitable dollars to church efforts and Ruritan baseball leagues.

I'm sure a certain percentage of people would respond to all this with, "Who cares? Farmers destroy the landscape anyway, so the fewer we have, the better. Reversion to wildness and wilderness is far better than producing food and fiber. At least the land won't be harmed."

Frankly, that kind of sentiment is only uttered by someone who hasn't missed a meal for awhile. The food you're enjoying must be produced somewhere, somehow, by someone. Is it righteous to move the damage caused by some kinds of farming from here to somewhere else? China perhaps? Africa? If we can't see the degradation our diet causes, does that make us feel better? Regardless of land use, farming and food production requires an eclectic skill set we're in danger of losing. The land transfer issues may not be as important as the information transfer issues. No civilization has ever survived an inability to feed itself. Period. Anyone who thinks the U.S. will be the first to do so is living in emotional and intellectual la-la land.

Yes, much of this land has been damaged. Much of that was from greed and much was from simple ignorance. But the fact that it was damaged in the past doesn't mean it must be damaged in the future. Assuming so gives in to a tragic and unnecessary pessimism. The whole local food tsunami, the nutrient-dense integrity food movement led by the Weston A. Price Foundation (WAPF), and the groundswell of interest in sustainable, regenerative agriculture are testament to how a society can learn, atone, and change.

Now I ask the question again, *Who* will farm this land? You see, this abandoned farmland is not public land; it's privately owned. In most areas of the country, privately-owned land forms the foundation of the local economy. The property taxes fund the schools. Production drives labor, saw mills, woodworkers, butcher shops, flour mills and hardware stores. When this land's productive capacity ceases, all the economic and cultural activity that flows from that value dries up as well.

The glaring reality is this: much of America's privately owned farmland is in short-term limbo. When I say short-term, I mean anything under 20 years. That's a blink of an eye on a landscape timeline. It cannot stay privately owned *and* abandoned, at least for the long term. Land cannot be in limbo for very long. It must be transferred out of the current owner's control to someone or something else. Most will stay in private hands. But more and more, the private owners who control it are independently wealthy and view the farmland as personal recreational areas. While this may be okay for publicly-owned land, it doesn't work long-term for privately-owned land.

The old status symbol was to buy a piece of the country and farm it, the proverbial gentleman's farm. The new status symbol is to own a farm and NOT farm it, the prestigious private park. But that just ties the land up in the hands of buffoons who don't know the cost and requirements of keeping up a farm.

When we say the word farm, what we really mean is something a farmer has done to the land. We don't think primarily of trees, soil, and grass. We think of barns, tractors, cows, corn, and fences. Without the farmer's effort, the farm ceases to be a farm. It's simply a privately-owned piece of wilderness. Unless we figure out a way to retain active farmers, all the farmland preservation efforts will simply give us high falootin' private wilderness which does nothing to employ people, feed people, or contribute tangibly to society.

I don't share the notion that farmers inherently damage ecosystems. If farming inherently damages the landscape, then we'd better eliminate farmers. Believing in the inherent destruction idea promotes environmentalism by abandonment, which sees nature as too pure, too perfect, to be adulterated by human touch. To many radical environmentalists, if all humans died tomorrow it would be cause for planetary celebration. I find that a hard option to sell. When I look in the mirror and ask why I have this huge brain and opposing thumbs, is it to inherently damage the earth?

No. The reason for this blessed endowment is to learn and then practice how to massage our ecological womb to stimulate more efficient solar conversion into decomposable biomass to build more soil than nature could in a static state. Now *that's* a mouthful. But it is nonetheless a mandate on farmers to heal landscapes through the act of farming. While most of us know about farms that have caused tremendous damage, we also know about others that have built soil and restored springs of water. Human activity is the most efficacious destroyer *and* healer. For more on this, read my book *The Sheer Ecstasy of Being a Lunatic Farmer.* It'll give you all the details, with tons of humor.

One final note on this discussion about farmers destroying the earth: we need to include the culpability of a society that abandoned domestic culinary arts. Americans during the 1950s and 1960s quickly embraced TV dinners, Velveeta cheese, soda pop and mega-corporate food processing, which dramatically altered the household food-scape as rapidly and profoundly as farmers who adopted feedlots, factory houses, and chemicals. Today, the fast food and snack food industries — as well as the headlong rush into genetically modified organisms (GMOs) — convict both producer *and* consumer as accomplices in wrong-headedness.

Voters who decided, through duly elected leaders, to subsidize six crops created a lopsided pricing structure that incentivized feed-lot beef, grain-fed dairy cows, high fructose corn syrup and Pop-Tarts in the marketplace. Promising convenience and freedom from menu preparation and planning, the big-food industry more than happily filled the opportunity created by vacated kitchens. That people could so quickly embrace the vacuous notion that ultra-pasteurized, shelf-stable, destroyed-then-pseudo-enriched foods could actually yield wellness should give us all pause. The Nature Conservancy member who does not resist the grandchildren's incessant pleas for a Happy Meal aids and abets this whole damaging food and farming system. Goodness, letting children watch TV bombarding them with government and corporate mischief implants in young minds hunger and thirst for nutrient depleted, soil eroding, sickness-engendering pseudo-food. I think there's enough blame to go around, don't you? Me included.

Let's agree that we need food, that farms are generally where we get the food, and that farms can't exist without farmers. Let's even agree that we need good farmers, not bad farmers. We could even agree that bad farmers and good farmers exist, and we need more good farmers, not fewer. Losing good farmers, the food they produce, and the land they massage, is not some theoretical problem to be solved by nerds in an academic focus group. The loss is serious, immediate, and pressing. Its reality underscores every page of this book.

Anytime a business or economic sector drops below age 35 in average worker age, it's in decline. I don't know how long it's been since the average American farmer was 35, but it's been decades. What happened between Thomas Jefferson's vision of intellectual agrarians and today? What happened to farming's allure? Farmers used to be held in high esteem, voted into public office, honored by the community.

Today, farmers are the butt of red-neck jokes and hillbilly stereotypes. School guidance counselors steer promising, intelligent young people away from farming and towards just about anything else. As a society, we've relegated farmland stewardship and the foundation of our food system to society's bottom socio-economic tier. When is the last time you heard a cluster of moms excitedly discussing the aspirations of their child prodigies envy the one mom who proudly announced: "My Jane wants to be a farmer." On the contrary, the cluster would mourn for that poor underperforming child. I look for the day when such an announcement is met with an affirming chorus of, "Wow, cool, awesome!"

Unfortunately, our society affords farmers little respect. Maybe the day of the celebrity chef will eventually morph into the day of the celebrity farmer. I look forward to the day when parents encourage their children to be farmers, rather than discouraging them from choosing such a bottom-tier vocation. No parent is more secure than one who has a farmer for a child. Farms offer a haven of rest and retreat, food, and care. Farmers are used to babysitting lambs and calves and chicks. Farms are well suited to multi-generational living, allowing grandma and grandpa a useful role in farm life and connection to the next generation even as their adult children help keep them away from the anonymity of the nursing home.

When I returned to the farm full time my friends and mentors mourned the squandering of brain power. What a tragedy, that I would throw away my academic prowess for a lowly farmer's life. I'll never forget the high school guidance counselor who went into veritable apoplectic seizures when I demanded to take typing rather than physics. Then when I told her I really wanted to farm for a career, we had to call paramedics to revive her. Not really, but you get the picture. It was not a pretty sight. I never counseled with her again.

Even with all this societal prejudice, however, I'm convinced the main reason young people don't aspire to farming is economic. Be real; young people follow the money. Money follows opportunity. Most farmers aren't earning a good living. As a result, many if not most farmers don't want their children to become farmers. "I want a better life for you," they admonish.

Much of this loss in economic opportunity is a direct result of consumer disconnectedness and a profound lack of participation in preparing, packaging, and preserving food in the home. While some may blame seductive advertising campaigns, in the end all of us are responsible for our choices. Otherwise, we dismiss irresponsibility with the excuse: "I'm just a victim." Nobody made Americans abdicate our kitchens. Nobody required that we become dependent on industrial food conglomerates and supermarkets. We chose to change. When domestic culinary arts are subcontracted to Kraft Foods, Cargill, and Archer Daniels Midland, not to mention McDonald's and Taco Bell, the portion of retail food dollar going into the farmer's pocket dwindles. In the 1950s farmers received nearly 40 percent of the consumer dollar. Today that figure averages 9 percent and it's continuing to trend down. The old saw about there being more money in the cardboard Wheaties box than in the wheat it contains is actually based on truth. If this is the expected template, who would bother to farm?

As families abdicate their kitchens for shopping malls, celebrity entertainment, and recreation, they buy processed and prepared food that fundamentally changes the economic landscape in which farmers operate. Instead of selling into a short food custody chain, they sell into a long one. Between field and fork a host of aggregators, wholesalers, speculators, laboratories, packagers, distributors, processors, marketers, warehouses, attorneys, and cash registers dip into the farmer's former profit margin.

This is why direct purchasing on the part of eaters is as important as direct selling on the part of farmers. Localization and shortening that chain of custody does more to create economic opportunities for aspiring farmers than any other thing. A permutation on this theme is bio-regional aggregation whereby a group of farmers dedicated to a protocol sell under one brand name but maintain a regional marketing persona.

Electronic aggregation is coming on strong now and promises to create economies of scale in local distribution. These are virtual farmers markets, using Internet shopping cart software to consolidate local offerings for consumers who pick up at a designated drop point. This takes some of the marketing pressure off the farmers (who are notoriously bad marketers), allows them to stay on the farm instead of spending a day at the farmers' market, and offers buyers a convenient pickup time for the entire local inventory. Buyers aren't limited to what is only available at a single point in time on farmers' market days.

As I talk with hundreds of young people each year, I'm convinced that if farming were seen as an economically viable vocation, it would immediately lose its dust-and-drudgery stigma. We're already seeing accelerating interest in farming among young people, but we'll get to that in a minute.

The single biggest element of this economic hurdle is entry level capitalization. The phrase "start farming" is rife with assumptions about buying land, buildings, and equipment. Expensive land, buildings, and equipment, I might add. This impediment raises a barrier of entry, keeping young people from being readily able to get into farming. Are you ready for a revelation? When young people can't get in, the old people can't get out.

If farmland can't get transferred to the next generation, old people are stuck with it. Most don't want to sell it outside the family. Most don't want to see it abandoned. But barring some miracle that defies societal prejudice, economic assumptions, and human resources, the land is in limbo until death or some other catastrophe jars it into other hands.

Interestingly, I had dinner recently with a lady involved with a large farmland preservation trust. She lamented that all these farms were dying for lack of young people. Her constituents were growing older by the day, unable to care for their barns, fences, and fields. Caretaking responsibilities such as regular mowing to keep the fields open, repairing fences or fixing an outbuilding's leaky roof were becoming prohibitively expensive. The farm didn't generate any income to compensate for these expensive repairs.

Her cry was clear. "*Who* is going to take these farms? Where are the young people to care for this land?" I began laughing because that very day was our deadline for internship applications for the following summer. We had eight spots and 300 inquiries, resulting in 203 completed applications. Just having to answer ten simple questions culled a third of the initial inquiries. To make sure you appreciate the math, this means we could only take four percent of the applicants. Yale and Harvard take a higher percentage than that.

She caught her breath and asked me incredulously, "You mean, that many young people really want to be farmers?"

"Yes," I assured her. We looked at each other for a moment as this revelation simmered in her psyche.

Stumbling through the obvious next question, she stammered, "Well, well, how do we get these two groups together?" And that, dear friends, is the theme of this book. *That* is the need of the hour.

The problem is that we need a seamless succession model to allow these two groups to work together. You can't blame the current farm owner for distrusting a young whippersnapper. Old folks have never trusted young folks. Young folks have to earn that trust. That requires a track record. It takes time. It takes experience.

Just as in other professions, aspiring farmers need to come with a farming resume. But if young people can't get in, how are they supposed to develop a resume? "Two years experience required," is the age-old dilemma facing new job applicants. Exasperated young people wonder how they're supposed to get the experience if employers won't hire inexperienced people. Employers, on the other hand, don't want to be wet nurses and babysitters. And they sure don't want to turn over their hard earned equity to an untested youngster.

Where are the young people going to be vetted? This is exasperating for both parties. The farmer looks at the line-up of potentials and tries to look behind the eyes to see the necessary character traits for success. The young farm candidate has no clue if this farmer has the mentality of a slave master or the patience of Job.

Neither party comes to the table with enough information to make a good decision. It's a lose-lose stand-off, and the tragic reality of our day is that literally thousands of people, on both sides of this table, are out there wandering around in a frustrated state of despair. I believe that for all the time spent talking about the farm bill, foodie dreams, and environmentalism, the elephant in the room is this awkward dance in which the partners never meet. The actual people who will actually do the work never join hands, and that derails all the good agendas conceived by well-meaning dreamers. Day turns into night, and night into day, and the two people continue in their world, desperate for a partner, but not knowing where the other is. This is far worse than not knowing which partner to choose. It's a case of not even knowing where to find a partner.

Many years ago I dabbled in trying to facilitate matches. Big mistake. If it doesn't work out, both the farm owner and the young person blame me. When I say I receive stacks of letters requesting young people to take over farms, invariably young people write asking for this list. No way. I'm not a matchmaker. That's like putting my head in a meat grinder. The point I'm making is that the need exists. The opportunity *is* out there.

Young people don't need to buy a farm, tractors, and buildings. In some ways, I think it's easier to become a full time farmer today than at any time in American history. A burgeoning interest in local food among the populace creates unprecedented retail food commerce opportunities. Many people are looking for local food with integrity and willing to invest in it. Never have I seen such a hunger for the kind of food that heals farmland while also inspiring eaters and reviving local economies. Pastured-based livestock. Compost-fertilized produce. Fresh, locally-produced anything. The variety is huge, from honey to lard to homemade relish.

At Polyface, our nearly 20-year experience with interns and apprentices proves my thesis that land is available, sometimes at no cost. While there are certainly pockets that offer more opportunity than others (such as upstate New York or mid-Missouri) to my knowledge we've never had a graduate of our intern program unable to find land. More often than not, they have to select among numerous land offers.

Meanwhile, most young people aspiring to be farmers complain about not being able to find an opportunity. What's the difference? Why is one group regularly turning away opportunities while so many struggle to find a way in? Very simply, a credible internship experience. It makes all the difference.

Young people need a jumping off point as much as the older farmers need a partner. While I have no desire to match up the actual participants individually, I find great pleasure and emotional reward in facilitating a platform that allows the matchmaking to occur. Young people need the self-confidence and resume to access the opportunities. The older farmers needing a partner rightfully demand a resume that's more substantial than smiles and dreams.

Established farmers want and deserve a vetted young person, someone with a track record. This nexus of need and aspiration is what interning is all about. I believe this partnering is the most important aspect of our food system today. If we don't figure out a way to get these two together, important and valuable farmland will languish unproductively and our food system will deteriorate. Rural communities will deteriorate as well. This is much broader than family transfer, although that is one partnering possibility, and arguably the most historically common. Today, we need to embrace non-family multi-generational agrarian partnerships as the key for this land and informational transfer.

Multi-generational farm partnerships create vibrant farm landscapes for all of our grandchildren, both urban and rural. Ultimately, this leaves a legacy of food and farming that is more secure and stewardship oriented. The future demands it. Tomorrow's generation deserves it. Now let's make it happen.

Chapter 2

The Field Classroom

F ields of farmers are not going to come from existing farms. By and large they won't even come from the country. Interestingly, most of the passion for farming is coming from urban and suburban areas.

A major anthropological and sociological correction is occurring, much like when house maintenance is neglected for a long time. The falling gutters, leaky roof, rotten window sills and collapsed front steps need a major overhaul to compensate for years of neglect. What could have been many minor repairs is now as expensive as building the house in the first place. The house of farming has been neglected for too long.

With rising awareness about soil depletion, urban food deserts, nutritional deficiencies and non-infectious health problems, young people, as they always have, are stepping into society's area of great need. We had the agrarian economy, then the industrial economy, then the information economy, and now we're entering the regenerative economy.

No longer can we pillage willy-nilly and expect things to be okay. No longer can we dump toxins, or throw things away. In the words of author David Orr, we're learning that there is no *away*. It's all here, as Thomas Friedman points out in *Hot, Flat, and Crowded.* Urban young people are more aware of these issues than anyone. Their interest in regenerative earth stewardship is a natural outgrowth of this new awareness. Thankfully, many of them are realizing that jumping in as self-directed entrepreneurs is preferable to lobbying for more government programs, regulations, and redirected subsidies.

From classic conservative to tree hugger liberal, a can-do attitude permeates the hearts and minds of many young people. For old geezers like me, it is a welcome wonder to behold! When I make presentations at colleges as part of an endowed lecture series and the students host a potluck with me, that's exciting. I can assure you no college class would have done that in the 1970s. These potlucks spring out of postage stamp gardens, pot gardens (not marijuana, but growing vegetables in flower pots — sheesh), Community Supported Agriculture shares (CSAs) and bartered arrangements with nearby farmers.

To see the light in the eyes of these young people, desperate to have a visceral connection to the land and be part of the solution is unspeakably gratifying. Their hearts are in the right place. They gather around me like doting fans at a rock concert. The refrain that dominates our discussions is about the next step. They ask, "Now what? I'm here in college, have $50,000 of debt and a degree in environmental sciences, but where do I go from here?"

To make sure I convey how broadly this desire permeates our culture, when I speak at conservative homeschooling conventions, it's the same thing. Families who have opted out of institutional education and conventional thinking, buoyed by the spiritual and emotional satisfaction of traveling a different road, are ready to embark on the next great possibility. Often, families who choose alternative paths soon yearn for additional ways to declare and *live* their independence. As night follows day, it seems, these families who embrace homeschooling soon find alternative health care procedures and begin patronizing quacks like chiropractors, naturopaths, homeopaths, and acupuncturists.

The next thing you know, they're finding midwives and doing home birth. Then a grain mill shows up in their kitchen and you see buckets of whole grains stashed about, milled on-site for fresh baking. Mmmmm. Then a garden appears out back, a beehive, and half a dozen chickens. Before you know it, a spinning wheel moves into the house and hand-knitted scarves and mittens appear where once Wal-Mart's "Made in China" accessories dominated.

This is as non-partisan a movement as anyone could imagine. Democrats and Republicans, Greenies and Libertarians, sit down together to participate in these living arts. Food and earth stewardship are truly societal connectors. You can do without cell phones, flat screen TVs and *People* magazine, but you can't do without food and fiber for very long. You'll be naked and hungry, and few us want to be there, especially when it gets cold.

I look out on this landscape, into the eyes of these bright-eyed young people, and the nagging question I feel is, *How* are these young aspiring farmers going to learn? They've embraced farming in their minds. They've embraced earth stewardship in their hearts. They're ready to jump, but how, and where, and when?

Different Models for Learning

Teresa and I homeschooled our two children, Daniel and Rachel, very early in the movement. Back then, we were still drawing the shades and hibernating during the day lest we be turned into the authorities for truancy violations. Fortunately, the Home School Legal Defense Association (HSLDA) came to the rescue for thousands of families across America facing the onslaught of education extortionists (just a more academic version of the Mafia) and created some wiggle room in the courtroom for this fledgling movement to germinate.

But embracing homeschooling early in the movement put us at a disadvantage. The extracurricular activities enjoyed in homeschooling support groups today, from soccer to debate, didn't exist in those days. Materials were scant. We bought used textbooks from the public school system. Networking was practically non-existent. Fortunately, even though we made plenty of mistakes, both of our children have excelled in adult life, so we must not have done too much damage.

Daniel never went for a high school GED. He didn't read until 10, when he was elected historian of his 4-H Club. The largest international youth leadership organization in the world,

4-H has local clubs throughout the U.S. Both Teresa and I actively participated in 4-H during our teen years. It is an excellent youth organization to offer social and developmental opportunities to homeschoolers. He came home that evening and made the grand announcement, "Well, I guess I'd better start reading if I'm going to do that job." Two weeks later he was reading. Later, he served as president of the state-wide Virginia 4-H Club with a college-bound cabinet that ribbed him about staying on the farm. He took it in stride and formulated a confident response: "Anything I need to know, I can find out. I don't need a college classroom to answer my questions." There you go.

Rachel read early and pushed herself academically. Coming along later than Daniel, by the time she had finished high school course work, Virginia had a ceremony for home school graduates. She had to name her school, and creatively picked "Head Held High School." Chip off the old block, I'd say. Then she went on to community college for a two-year associate business degree, earning straight As. The first test she took was in college.

Then she went on to The Art Institute of Charlotte for a two-year interior design degree. Again she excelled, achieving top portfolio in her group. No, I'm not biased; just proud. Today, she manages a retail store front for a guild of about 100 local artisans who sell collaboratively by consignment. She brings her artistry and design gifts to Polyface in numerous ways. Whenever we have a construction project, she designs it. She also puts together my PowerPoints and created her own Polyface post card and note card collection, with pictures from the farm. This spring she released Coloring Polyface, an accurate info-dense coloring book for ages 3-8.

Homeschooling has a long, rich history. People have been learning in non-institutional settings far longer than there's been compulsory public education, which started relatively recently. Really, only for a little more than a hundred years has going to school been the norm. Before that, for millennia, home-based tutors and homeschooling were the norm, producing the majority of the great thinkers and contributors to history and society — and the regular Joes, too. Yet in spite of this enormous precedent, at some point in recent history the homeschool approach became suspect. But today, the whole issue of non-institutional learning is rife with emotion. As soon as someone embraces non-institutional learning, the entire institutional apologetic raises its hackles. School teachers, college professors, administrators, and boards of education. Oh my, this is really a Pandora's box. I realize the quicksand I'm approaching by dipping my toe in this issue. But it must be broached, because it's the foundation for internships.

Much of the polarization on this topic comes from shortcomings on both sides. For the sake of discussion, I'd like to shift the word institutional to classroom. Interning is an institution too, so let's be fair and describe the two models as classroom learning versus experiential learning. Ah, I can hear the voices booming back from these pages, "But our classroom *is* experiential!"

And therein lies the crux of the debate. From Montessori schools to Waldorf education to a host of independent-minded creative models, a move toward hybridizing classrooms with experiences, academics with encounters, is on-going. It's wonderful.

One of the early critical terms used by the homeschooling movement to voice concern with standard education was disassociated learning. It described the need to associate academic endeavors with hands-on experiences. Essentially to test theory with practice. For example, when children learn fractions as a result of helping measure boards to build a rabbit hutch — as in 5 ft. and 1/2 inch — suddenly fractions take on new meaning. A light comes on in the student's eyes. She understands how what she's learning relates to the world.

On the opposite side is confusion, and boredom. When students say, "Why do we have to learn this?" that's a dead ringer for a disassociated lesson. For too long, classroom teachers have dismissed this question as being irrelevant. But it *is* relevant, because the why creates the motivation, the interest.

As a result, through the homeschooling model the educational pendulum swung toward corrective action with experiential learning. Indeed, when we started John Holt's unschooling was the darling of the liberal branch of homeschooling. The conservative branch, led by Raymond and Dorothy Moore through wide exposure on James Dobson's *Focus on the Family* program, charted a more formal approach. Because formal debate played such a formative role in my own educational experiences, I tried early on to get our home school support group to let me coach, and launch, a homeschool debate team. Parents were appalled at the thought of competition. "Competition is evil!" they told me, explaining that was one of the reasons why they left institutional education. The interesting thing about cultural pendulums is that once they reach the end of their swing and begin coming back, they don't stop in the center. They keep on swinging past balance and go all the way the other way. These swings can take years for people to see. Today, our local homeschool support group sports a debate team, soccer team, basketball team. My, my, how things change.

For example, our own American culture went from an agrarian homesteading, ruggedly self-reliant and pioneer-savvy one in say, 1900, to an industrial, anti-self-reliant mindset by 1960. Home canning and hog killin' was replaced by TV dinners and infant formula. Breastfeeding was widely considered barbaric and Neanderthal — no respecting woman would breastfeed her baby. Then the pendulum swung correctively into the beaded, bearded, braless Woodstock hippie generation, embracing environmentalism, communes and even La Leche League and Lamaze classes. Who would have thunk?

In my own lifetime, I've watched education go from phonetics to Dick and Jane's sight-see back to phonetics. From single classrooms to open classrooms to team modules to homeschooling.

The pendulum seldom sits on balance. It's always moving from one side to the other. Where that pendulum is when we enter life often determines, for the rest of our lives, our social and political perceptions.

Violently concerned about the path we saw in public education, Teresa and I were committed to an alternative but had not yet heard about homeschooling until after Daniel was born. We embraced the notion, in its infancy, catching the pendulum perhaps in its over-reactive state.

At any rate, as an onlooker into professional education, my sense is that the classroom tends toward disassociated academics and the experiential tends toward a non-academic approach. Both sides would be stronger by adopting techniques from the other. Classrooms are much stronger when they incorporate far more experiential and self-directed learning. The opposite has sometimes been called unschooling. While I deeply appreciate the idea behind unschooling, the tendency, in my observation, is to cavalierly dismiss academics altogether. And it is this weakness that raises the ire of the classroom apologists.

The pendulum is not currently in the center on either of these models. I have no problem with classroom settings. My problem is when they're considered more important than experience. Since most American classroom settings tend toward bureaucratically-directed, top-down models, independent-minded libertarians like me tend to bash the whole classroom idea as being constrictive. That's my weakness and I'm being as personally transparent as I can be, revealing that I too have my preferences, biases, and opinions. But I'm keen to connect with everyone in this discussion and find common ground where we can.

I'm going somewhere with you, but I need us both to understand something. Advocates for conventional classrooms have a blind spot: the need for more experiential visceral hands-on learning. Advocates for unschooling have a blind spot: the need for more academic challenge and excellence. Given my background, I'm being honest about my pendulum's off-balance tilt toward un-schooling. I doubt that any of us has a pendulum hanging directly in balance.

With that in mind, let's go on to have a discussion about internships.

I want us to think of internships not as an alternative to academic excellence, but as a necessary component of academic excellence. If we use the term supplement, that makes the internship sound less important. The experiential doesn't supplement the academic, but is equally important. Kind of like the brain can't function without the body, or the body without the brain. They work best in tandem with neither thinking it's dominant or better.

Internships exist in nearly all fields and all over the world. Sometimes they are called co-op programs. Sometimes they are called experiential immersion, like in German college programs, which require 6 months per graduate. Some internships are paid and some are not; some are longer and some only a few weeks. Some are formal and some informal. The beauty of internships is that they allow a person real-life access to an intended vocation without the full responsibilities of actual employment.

Interns put in a full work day, but their status builds forgiveness and shepherding into the relationship with bosses. Too, businesses usually try to expose interns to many different situations within the work place, trying to create a well-rounded experience. While the intern performance levels receive some slack, great performers usually receive requests for full-time employment. The opposite is also true: poor ones can be relegated to being shunned. While interns learn something about their future vocation, mentors learn something about their future workplace partners.

Internship offers a space with some wiggle room where vocational novices and candidates may dip their feet in the water of workplace reality to see if it's a pool they really want to swim in. By the same token, it's an opportunity for business people to try out candidates without all the trappings of protection required in formal workplace arrangements. Historically, interning was a way for someone to learn a trade with very little capital; a way literally for children to grow up with a marketable craft.

Farm internships usually span the busy summer season. Interns normally live together, sometimes in extremely rustic conditions, for the sheer privilege of getting their hands dirty participating in farming activities. While many just want the short-term experience to pad their memories and resume, most have a deep-seated desire to actually test the viability of a farming future. Internships are full farm immersions wherein young people engage in every facet of the farm, from marketing to pulling weeds to slaughtering chickens. Sweating together in hands-on physical work creates lifetime friendships, both among the interns and with the mentoring farm family. Interns are not coddled, but enter the farm operation on the lowest rung and advance according to aptitude.

A Pencil Wielding Lunatic Farmer

I can best illustrate my point by telling my own story. My liberal arts college education was interesting, but what I really took away from the experience was four years of intercollegiate debate competition. Dad used to say I majored in debate and minored in everything else. I didn't make excellent grades; I lived for debate.

Many people have said that my writing must have been developed by virtue of my English major. No way. As early as third grade I would come home from school and just for fun sit at a makeshift desk in the dining room and write fiction with a number two pencil in a lined spiral notebook. From my earliest entry on up through college I was winning local essay contests. I can't remember entering a writing contest I didn't win.

I'm not bragging; just pointing out that my college English degree had nothing to do with my love for or competence in writing. I was blessed with a couple of excellent high school English teachers who certainly massaged my mechanics. My junior and senior years in high school found me waking up at 4 a.m. on Saturdays to arrive at the Curb Market food vendor's stand in Staunton, Virginia. Home by noon, unpack, and then go into my evening stint as receptionist at the local daily newspaper. Writing obituaries, police reports and assorted blah news items under the tutelage of a colorful newsroom staff of reporters, photographers, and editors honed my clarity, brevity, and accuracy. I've been writing and farming as long as I can remember.

In high school, I competed in forensics — interscholastic debate — for five years, and played leading roles in numerous dramatic productions. Nobody could be more amazed than I am that people want to hear a lunatic farmer deliver a dramatic monologue about the ecstasy of farming. Perhaps more importantly, who would guess that I would take these storytelling and writing skills to a new level of importance as I persuaded people to try pastured chicken and salad bar beef? No question, many times I wake up feeling like Cinderella, and the magic never wears off. How many people enjoy the luxury of turning all the greatest loves of their life into a profitable avocation? These books about having a job to earn money so you can do what you really want to do in your free time--I don't get it. I'd do this for free if I didn't have to pay taxes.

By the time I began my college English major, I had clocked thousands of hours writing and speaking. In college, at Bob Jones University, I was elected vice president of the prestigious Writers' Forum, a group of professors and students interested in literature and writing. Ever the contentious rebel, I had this sneaking suspicion that all those English papers about what an author was thinking as he wrote were just a bunch of foolishness.

I believed good writers were just good writers. They practiced a lot, had a passion burning to be put on paper, and followed patterns they'd seen effectively utilized by masters of the art.

As vice president, I was responsible for the club's programming. Against the better judgment of the other officers, I proposed that we officers submit pieces we'd written for the academic scrutiny and judgment of the members. The pieces were submitted under the guise that they were written by masters in literature, not by students. Sure enough, when my piece came up, the heady assumptions bandied about by the group revealed completely over-the-top interpretations that were entirely off-base.

"Notice how the author is using food to set the stage for describing the evening sunset." I assure you, I never thought about food when writing it. During the whole meeting, those of us who had submitted pieces had to practically run from the room to keep from giving ourselves away. I was dying laughing on the inside. All I had done was whip out a one page slice-of-life without much thought at all. In fact, if I remember correctly I composed it on a typewriter (yes, one of those things) on the fly, no corrections.

Talk about a rebel. To me, that showed the fallacy of all those literature papers that purport to discern what's going through an author's mind when writing. No; good writing flows from the heart and is just good writing, good communicating. Now, as to the meaning of a piece, that's important to discern. But all the projections underlying it — a bunch of hogwash. You can imagine that this exercise, once it was discovered, didn't endear me to the English faculty.

I did have a great college experience. But in the end, it was a lot of money to get a few great friends and compete on a debate team. What if I hadn't gone? What if I had kept the money? I'm not saying I wish I hadn't gone to college; don't read more into this than my brainstorming ruminations allow. At the time, I thought the only way to become a full time farmer was to get a degree and earn money somewhere else. This was the Watergate era, and as a budding journalist from a libertarian family, I thought I'd expose the next great error in Washington, write a best-seller, than retire to the farm like Henry David Thoreau. That was my ticket to the farm.

The summer after graduating high school, I shut down my Curb Market stand, which by this time had grown to quite a business. My chickens, eggs, and vegetables, plus the family's dairy products, beef, and pork, were gaining a foothold. This was 1975, well before most people had heard of the O word (organic). Pastured beef and chicken were not in anyone's lexicon. But I had a special dispensation to sell uninspected products at this market through an agreement between the Virginia Extension Service and the Food Safety and Inspection Service.

By joining 4-H, I was in the government system, so to speak, and therefore vetted as a bona fide participant in safe food. This market, precursor to today's Staunton Farmers Market, was a holdover from the depression era. It had dwindled to two elderly matrons and me. By the time I got home from college, they had closed up. But while the three of us were there, the ladies taught me marketing and how to handle customers, how to present myself and showcase the products. And I like to think that I injected youthful vitality into their lives.

But the question will always nag me, *What* if I had stayed home, continuing to farm and work the Curb Market stand? I'm now the only living vendor from the Staunton Curb Market — quite a distinction. Interesting that I'll hold it until my death — this is one record nobody can surpass. Everyone else has already passed.

Would those four years of growing my farm business and enjoying the marketing freedom of the Curb Market have been as advantageous as college? I'll never know, but it's a valid question. I wouldn't have had the same experiences, but I would have had different experiences. I would have been grandfathered into whatever showdown would have occurred when one or both of those ladies decided to quit. That may have fundamentally altered my farming and writing trajectory. On the other hand, I may not have been as confident facing the world as I was with a college degree and all its associated experiences.

I do know that for regulators to shut us down from our unregulated pastured dairy, meat, and poultry products would have been a real battle. We had many customers and the market was a unique Staunton institution. Many retired vendors still had great appreciation for the market's continued existence. History can't be rewritten, but I've spent many hours musing over the strategic timing of my experiences there. Our farm was literally only 5 years ahead of its time. So close, so close to the food revolution that would grow out of the budding hippie back-to-the-land movement germinating in the mid-1970s. What an exciting time, and we were right on the cusp. Could we have survived until the birth of the O word? Fascinating what-ifs, aren't they?

By 1979, when I graduated college, health food stores were appearing. Articles about food faddists appeared in newspapers. The hippies had spoken. The Vietnam war was over. If I had hung on during that time, who knows what would have happened? Would I be a different person? Yes. Better or worse? Who knows? We can take many paths to a destination. I'm extremely glad I went to college and certainly don't consider it a waste of time. But it was not the only path to where I am today.

Since Polyface is only a couple of hours away from Virginia Tech, I'm sure it seems only natural for me to have gone there. That's probably why people ask. Assuming that I'm a proud Hokie, they probably also think that this is what has lead to my successful approach to farming and the vitality of my farm as a business. But like I said earlier, experience outside of the classroom — hands-on, experiential learning — is another path to success in any field, farming included. I didn't major in agriculture. And I didn't go to Virginia Tech. I learned farming over a lifetime, with Dad and Mom and other farmers as constant teachers and guides. In fact, when I was in 4-H leadership, professors at Virginia Tech did try hard to recruit me, but the confinement poultry operations I saw on these trips did not impress me. I was on a different path, and it has panned out quite well.

Frankly, I'm not altogether sure getting a degree in agriculture is the best thing anyway, regardless of its paradigm. I've had animal science and agriculture majors from land-grant colleges come to the farm for tours or short seminars. At the conclusion of one a bright young fellow said to me, "I've learned more here in four hours than I did in four years at Virginia Tech." He went on to say that in college he memorized the Latin names of every bone in a cow's skeleton, but nobody explained the essence of cow to him. Well *that's* my specialty — I'm famous for extolling the "cowness of cows."

Dear folks, I'm not trying to disparage Virginia Tech. It could have been any land-grant school and the story would likely be the same.

Another fellow with an agriculture degree said he spent ten years *un*learning what he learned in college. This isn't uncommon. How can this be? I could step off the edge here and go into a tirade about agriculture colleges being puppets for industrial farming corporations. But I won't.

The problem with the agriculture schools is the same problem I encountered in my college writing club. Academic institutions tend toward theory and not practice. Because they exist in a sterile environment that rewards smarts over practice, they develop a haughtiness. What the writing group should've concentrated on was *not* dissecting what may or may not have been in a writer's head, but how to excite passionate conviction in the hearts and souls of young communicators.

Before these aspiring writers put pen to paper, they needed to answer the question, "What is such a compelling issue in our world that it would drive me to work night and day?" Or perhaps this: "What values would I die for?" All of us can use good editing. How about submitting to editing? Who cares what an author was thinking? All we can know is what she said. That students spend thousands of dollars for the privilege of writing papers analyzing what they believe an author was thinking strikes me as a serious waste of money and brainpower. It's better to examine ideas and develop convictions sacred enough and compelling enough to consume our lives.

Field Studies

One of my favorite activities is to go out and lie down in the field with a herd of cows. If you've never done it, I highly recommend it. You can't go lie down comfortably in a field of pigs because they're omnivores. Do I need to explain that? Okay, the pigs will come up and begin nibbling your shoes. Shortly they'll nibble off your ears — I mean nibble off your ears. They're omnivores. In a couple of hours, they'll get to your spleen and liver. I always warn school kids who come for farm tours to keep moving when they walk through the pigs. If you quit moving, they'll quite literally have you for lunch.

Cows, being herbivores, won't eat you. In the cool of a summer evening, few things are as soothing as lying down amidst a herd of cows. Especially our cows, which have normally been moved into a new paddock a couple of hours before and are now bursting with full stomachs. Their left side is pooched out to the breaking point because they've just ingested up to 150 pounds of grass. How's that for a normal supper?

With the sun casting lengthening shadows right before it slips behind the western hills, the cool sod envelopes me as I nestle into it. Because our cows are in a tight mob (you can read about this in my book *Salad Bar Beef*) it doesn't take long for a cow to amble over to this prone human in the grass. The thousand-pound bovine approaches cautiously, not sure what to make of this strange object in an otherwise savory salad of forage. Slobber dripping from her chin and a tinge of green lining her lips like misplaced mint-flavored Chapstick, she stretches out her neck, sniffing. If you've never been nuzzled by the muzzle of a cow, it's hard to appreciate just how big a bovine nose really is. It's big. You can scarcely put your hand over it.

She stretches, sniffing, sniffing, craning her head forward to touch the toe of my leather work shoe. Does she know about leather? I don't think so. I think my feet smell more like me than cow. Actually, they probably smell like grass, but she doesn't bite. She just inhales. She runs her nose, ever so gently, over my shoe, like a blind person would run gentle fingers over it to see it through tactile interface. I lie motionless.

Uncomfortable with her neck craning, she finally takes another step forward. Ah, that's better. She can now push my foot around a little bit, first with her nose, then the bottom of her mouth. Her wet nose has now dampened my shoe. Assured that I won't hurt or scare her, she suddenly reaches out a nine-inch tongue to give my pant leg a lick. I think it's sweat, perhaps a bit of saltiness, but something surely attracts these cows to begin licking the human body.

By this time, another cow, attracted by the sounds of the first one, eases over. Together they explore my legs. Since I can't see behind my head, I'm first aware of a third cow approaching from behind by her soft breath on my forehead. She licks my hat and it slips askew. With head fully exposed, cow number three licks my forehead. All the activity has attracted several more cows, who have approached from both sides.

Within 10 minutes of lying down, I'm surround by no less than ten beasts weighing a total of 10,000 pounds, all nudging, licking, sniffing, and snoodling. Their tenderness is palpable. I've done this countless times and never had one step on me or bite. The key is to lie motionless, in complete submission to their interest. I have to gain their trust by not flinching — that would scare them away. Gradually the first comers will back off to allow a second rank of body-sniffers.

This can go on for some time until the last one has had her encounter and ambles off to lounge and chew cud. Often one will sneeze or snort, which scares the others and they all jump away, jittery for a moment. Then they all realize it was one of their own that snorted and with seeming embarrassment they come back to their sniffing and licking. If I'm wearing a straw hat, they'll usually try to eat it, so I have to lie on it or hold it tight.

Once they're all gathered around, I can shift body position slowly, deliberately. They will abide movement, but it must be extremely non-threatening. Learning their toleration level is an art.

Colleges don't teach you this. To an academic teaching Latin bone names, this warm cow fuzziness is ludicrous. But I've learned that it's the real beginning of understanding. It is practice. It is encounter.

What excites your spirit more? To have the kind of experience

I just described, or to recite the Latin names of all the bones in a bovine skeleton?

I'd bet that if cowboys had been running the USDA for the past four decades, official U.S. bovine feeding policy wouldn't have promoted feeding dead cows to cows. But because the USDA and the beef cattle industry isn't run by cowboys, the academically credentialed experts adopted a policy that ultimately created bovine spongiform encephalopathy (mad cow disease).

If we're going to populate our farms with good farmers, we need to encourage encounters like the one I just described. Engaging with life on a practical level, a visceral level, puts ethics and common sense around the egg-headed theories of professors. Most professors in agriculture programs grew up on farms but couldn't figure out how to make a living on a farm. They did the next best thing. They got a PhD from a land grant college and then spent their lives telling other farmers how to farm. Why do farmers listen to these guys?

I put a lot more credence in the advice of a true-blue successful farmer than the pontifications of a college professor. When I look for advice, I look for someone who successfully practices what I want to practice. That being said, there are certainly a few good college professors. But the system gravitates toward foolish advice. America's farmscape is littered with bad advice from university experts.

One friend facing bankruptcy with his hog farm lamented to me that, "The frustrating thing is I did everything the university experts told me to do." He learned the hard way that we are as responsible for picking our teachers as teachers are responsible for what they teach.

Perhaps we could say that farming is a bit like riding a bicycle. You can know how a bike works and read about its history. You can know how to take it apart and put it back together. You can read about gravity, centrifugal force, and balance. You can read biographies about great biking enthusiasts. But until you get on the bicycle and pedal, you can never appreciate the nuances, the reality of riding a bicycle. Without actually riding, you can never experience the exhilaration either.

It's one thing to know about something. It's quite another

to *really* know it, experientially. Imagine something as simple as pounding a nail. It sounds simple, doesn't it? At Polyface, we take interns who have never pounded nails. The skill and art of pounding a nail can't be taught in the classroom. A hammer head is rounded, not flat. That enables the pounder to use slight side pressure — or fore and aft pressure — to move a nail directionally. Sometimes you want to drive a nail at an angle. If the nail bends a little, you need to correct — at the exact bend in the head to give you the result you need.

Last summer we were building a fence and I had an intern nailing in the horizontal brace spikes. This keeps the brace from slipping out of its groove in the post. After bending over half a dozen nails, she said, exasperated, "Shoot. These stupid nails just won't go in."

I went over and pounded each of those nails, already horribly bent, the rest of the way in, right up to the head. She just stood there, shaking her head. Similarly, a few days ago we were pounding some rebar through some bridge planks into the ground at the abutment approach. We wanted to hold the first plank in place and that would hold the rest of them in place. An apprentice was pecking away at the rebar with a sledge hammer, tapping it in about a quarter inch per hit.

I asked for the sledge and drove it in with about six swats at three inches per lick. The apprentice laughed incredulously as I handed him the sledge back. "How do you do that?" he asked. "Experience," I said.

That's when I break out in my favorite country song by Toby Keith: "I'm not as good as I once was, but I'm as good once as I ever was." That's always good for a few laughs from the crew.

One of my favorites is teaching the fine points of post hole digging. The three tools for the job are post hole digger, digging iron, and shovel. With the post hole digger, you start out and establish the hole. If the ground is soft, you can dig the entire hole with that tool. The key is to thrust it down hard. Every single apprentice and intern we've ever had uses the post hole digger timidly, as if the tool is about to break. I have to show the technique of bringing it way up and jamming it down hard.

If the soil is a bit moist, it will adhere to the digger. I watch interns wrestling this clingy soil off, banging the digger around and swatting it on the ground. The proper technique is to slap the handles together on the release stroke, which shakes the dirt loose from the metal.

If the soil is a bit hard or you encounter rocks or roots, the digging bar comes into play. This is a rod about six feet long with a flat head like a big nail on one end and a sharp wedge on the other. The flat head is used to tamp the soil back in around the post and the wedge end is used for loosening hard soil, dislodging rocks, or cutting through roots. Believe it or not, it takes hours and hours to master the skillful use of this simple tool. Every situation is different. Sometimes aggressive force is best. Other times, gentle probing between two rocks is best.

Let's assume we finally have a hole ready to receive the post. Oh, I didn't mention the dirt pile placement. That's important. I wish I had a nickel for every time an intern puts the dirt on the downhill side of the hole. Of course, that makes sense when digging because it's easier to swing the post hole digger downhill than uphill. But the problem is that when you begin putting the dirt back in the hole around the post, it's hard to shovel it uphill.

By putting the dug-out dirt uphill of the hole, shoveling back in around the post is a breeze. But again, experience comes into play. Invariably, novices stand behind the pile of dirt to shovel it around the post. The correct place to stand is next to the hole so you can use the shovel side and just scrape the dirt down into the hole. It's considerably easier and more efficient. This body placement allows you to pull the dirt toward your body rather than awkwardly shoveling it away.

I hope this explanation dispels, once and for all, the notion that setting a post, using a shovel, post hole digger, and digging bar, are simple tasks. Mastery takes many, many hours. And when it comes to farming, that mastery can't be taught in a classroom. I was recently putting in a fence post with one of our apprentices, going over the finer points of efficiently handling a shovel. When we were done, she lamented: "You'd think that with something as simple as a shovel our bodies would have an intuitive sense about how to handle it. It just doesn't make sense that there would be so much learn about how to handle a shovel." Well said, indeed. This is why you're an apprentice.

It's prudent to point out that these basic farm skills have nothing to do with organic farming, sustainable agriculture, biodynamics, holistic resource management, permaculture or any other agricultural variation. These skills, which form the backbone of *any* agricultural pursuit, can be learned from any farmer anywhere. When looking for intern opportunities, don't assume that a fairly conventional farmer has nothing to teach.

On the contrary, most of the basic farming skills can be taught by any farmer. This isn't a book about organic farming. This is a book about germinating tomorrow's farmers; lots of them. Learning opportunities exist in far more places than you can imagine. People don't have to hold all ideas in common in order to learn from each other. Plenty of conventional farmers are better at backing a trailer than I am. Don't dismiss their abilities.

About half way through every season, I enjoy listening to our interns converse with visitors to the farm, explaining how we do things. The lights go on as they realize how much they know. It comes with daily practice. It's osmosis, really, by doing and living it.

I wish that all the money spent on college agriculture degrees could be pooled as a fund for wanna-be farmers. Anyone who thinks sitting in a classroom for four years accumulating $60,000 in debt, or spending that amount of money, to receive a degree in agriculture is better than spending four years interning on a successful farm simply isn't aware of what these two different models produce. One generates young people who look at every problem like a research question. The other produces young people with a git-er-done attitude who know how to walk it instead of just talk it.

Self-confidence for practical doing doesn't come from answering questions on a test. Self-confidence comes from pounding that nail in, driving it home with a sure hand because you really understand how to swing a hammer. Self-confidence comes from setting that post, thumping it firmly and feeling the sturdiness in its position.

We live in a culture where too many valuable — crucial — life skills have been lost to most people. If the country collapsed, who would you rather have on your team? An agriculture professor or someone who knew how to milk a cow, make cheese, and deliver baby calves?

For too long our culture has pushed college as the logical next step for all young adults, no matter their interests, abilities, or chosen profession. Yet for many, if not most, college is a huge waste of money at a critical, formative period of life. Rather than being a catalyst to the future, it's a noose around the neck. Instead of freeing young people to pursue their dreams, it enslaves them with indebtedness without necessarily equipping them in meaningful and useful ways. Then at age 40, tired of the Dilbert cubicle rat race, working for the man in some globalist fantasy, they suddenly realize they never pursued their real heart's dream. At that stage, many begin to wonder if it's too late, if they still can achieve their dreams.

Then they call me. "Could I come and follow you around for a week?" The answer is no, not for a week. You can't learn this in a week. It's too distracting. It doesn't fit with the schedule and lifestyle of real farm work. Although we've had older interns, most are either just out of high school or are college age. College definitely doesn't make a better intern. I'm not sure if it hurts but it certainly doesn't help.

I'm writing this chapter on a Qantas flight going from Melbourne to Perth. Today a beautiful young lady in a seminar I was conducting at Taranaki farm in Victoria begged me for advice on how to answer her parents who were insisting that she get a college degree. She has a heart for farming, and especially urban farming. She's already participated in some local food production projects and is ready to launch her full-time farming endeavor.

Her parents, like many, fear for her future without a college degree. Some of their fear comes from peer pressure. What will the family say? What will the neighbors say if our bright little Janey doesn't go to college? She's inquisitive, an avid reader, and gung-ho to go start farming. Why saddle her with a multi-thousand dollar debt?

Release her. Let her fly. Everything she needs to know she can learn outside the classroom. Self-taught folks abound. It's never been easier to teach yourself. In fact, she feels heady with information right now. What she needs is to get her hands in the dirt, to get some confidence growing things. That will drive her curiosity and experimentation.

On our farm, we deal with the same issue. Some parents are quite supportive of their children doing a farm internship. If a parent can get over the notion of their child becoming a farmer, they tend to get over the internship idea as somehow bad or a wrong turn educationally. Fortunately, the foodie movement that has created rock star chefs is now making room for rock star farmers. The stigma attached to farming may subside after all, but it's slow in coming. If this non-college approach scares you, then consider the internship as a supplement to college. Or take a break between sophomore and junior years to do an internship--hybridize your decision-making. Do an internship first, then if you feel unfulfilled, go on to college. One year more or less won't mean anything when you're 50.

Do you know what the average college cost could finance in a farm startup? Plenty of entrepreneurs have launched with a lot less capital than the average college education. We should quit wasting this time and money and instead do internships--or at least start with internships. The place where college degrees are important are in heavily regulated vocations like nursing, where government licenses mandate classroom regimens. Engineering and architecture often require licenses obtainable only through colleges. Until very recently, in Virginia you could sit for the Bar exam without college, simply by going through an extended formal apprenticeship program. In vocations where achievement doesn't require a license, interning offers a viable alternative.

Plenty of physical classrooms and Internet classes exist to satisfy academic yearnings. I don't understand why college degrees are honored more than entrepreneurial prowess. If our culture lauded innovative can-do accomplishment as much as college degrees, I believe we'd have a more empowered, skilled, and capable youth culture. Young people would feel like they have more options and our culture would be richer as a result.

Especially for Mentors

Chapter 3

Wages

Can the Polyface type of farming feed the world? I love that question. It's fun to answer and is generally asked with the sincerest of intentions. No problem. Of course it can. Is Polyface type food really just an elitist movement? That's the second most popular question and I enjoy that one too. Again, easy to answer. And, while it sometimes has a barbed spirit to it, is generally asked in sincerity. The price prejudice on local high quality food has many explanations, but elitism is certainly not one. I address this issue in several of my other books and won't delve into it here.

The question that raises my hackles though — that makes me wrestle with my inner being and say "down boy" — is this one: "Doesn't Polyface really succeed only because it exploits young people with uncompensated labor?" I always have to just stop, count to ten, take a deep breath, and pray for grace when answering this one. People have even accused me of turning young people into slaves. That's a powerful word and one I'm concerned about even using here, but I think I'd be unfair to my most vehement critics if I didn't acknowledge it.

What an absurd, insulting, and leading notion. As the rest of my book will show, an everyday internship and apprenticeship program like mine, and like so many others in myriad vocations, has nothing in common with the wretched institution of slavery. Anyone ridiculous enough to pose such a question sure seems to have an agenda of their own, though what it is, I don't know.

That well-fed, well-traveled, well-educated, well-read folks make this comparison all the time is patently absurd. Even if we housed these young people in squalid tents, fed them tepid water, paid them nothing, and fed them stale bread they have freely chosen to come and may freely leave any time they wish. Nothing, nothing, nothing about this smacks of slavery. Only a judgmental idiotic ignoramous would make such a charge. There, down boy.

How does a person with no apparent future chart a different course? Often it requires service to someone or something in order to create an emotional, educational, or economic nest egg to leverage into the next phase. How many economically disadvantaged young Americans use the military (called service, I might add) as a spring board to a better life? These young people haven't signed up to move chicken shelters; they've signed up to potentially die.

When these young people — by the hundreds, I might add — decide to apply to Polyface for a farm internship, they aren't entering slavery. They're doing it of their own volition. These are consenting adults, for crying out loud. They know exactly what they want, what to expect, and the terms. Nothing could more exemplify free will.

Well, then, isn't Polyface at least exploiting them?

I find it fascinating that American culture applauds young people who volunteer four years of their life to sit in classrooms to earn a diploma, with no guarantee of a job while accumulating an average of $80,000 in student loan debt. That is considered smart and savvy. But if a young person volunteers to come to Polyface for a four-month educational internship that costs nothing, in which they receive room and board, that the accusation can be freely bandied about that we are exploiting them.

The interns who do well at Polyface exit with numerous opportunities available to them. While we don't guarantee job placement, I don't know a single one that has floundered looking for compensated positions, either self-employed or employed by someone else. In their living quarters, the interns maintain a folder of requests that come across my desk almost daily. From Hawaii to Kenya to New England, farmers, non-profit organizations, and everything in between contact me looking for a Polyface intern to partner in sustainable farming enterprises. I wish we could fill every one, but the requests are too numerous. My heart breaks for all these requests because I know how helpful a Polyface intern would be to these endeavors.

Where are the young people ready to take on these culturally healing opportunities across the planet? Too many have been told that the only path for bright young people is sitting in a classroom paying tens of thousands of dollars into a system that, some would argue, exploits them. It takes four years of their life, indentures them through costly loans for half their working career or half their parents' retirement, and too often tells them the next step is another two years of the same. Yet few call this exploitation.

Education is expensive. Whether it's an institutional, academic environment or the proverbial school of hard knocks, education is expensive. It doesn't matter whether it's coaching to become good enough for the Olympics or a professor of environmental sciences or a farm mentor: education is expensive.

Education always has been expensive, and it always will be. Learning something takes time, money, and effort.

Before I go on, let me pounce on what I think is the core issue: learning farming or a trade seems different than learning academics. But, is it? When a tenured PhD professor has a cadre of students out collecting data on some research project that he in turn publishes in a scientific journal which compensates him through prestige and speaker fees, how is that different than a farmer who trades education for field work?

Academic institutions utilize students all the time, charging them for the privilege of doing the grunt work for projects that then redound to the prestige and economic strength of the institution — may we say, the business? Education is a business. Like it or not, it's a huge business. Many years ago when I was asked to do my first university lecture (I call them performances), I thought the college lecture circuit would run its course fairly quickly. I've now come to learn just how many colleges there are. I think I'll be long gone before I can get around to speaking at all of them.

38

Education is big business. "Publish or perish" drives this business. And the grunt work to make the business go is generally performed by bright-eyed-and-bushy-tailed, eager-beaver young people willing to pay $20,000 a year for the privilege of adding prestige, profit, and power to the institution and its faculty. This sounds mercenary, but it's the truth.

The only reason farmers like me are accused of exploiting young people in an internship program is because the trade between work and education is not as well hidden as it is in mainline institutions. On the farm, we're openly considered to be in business. But in spite of huge profits in college sports, fast-food courts in the student commons, industry partnerships, grant funding, the medical research gravy train, and a steady stream of sweatshirt and bumper sticker sales, universities are somehow not considered to be in business. They are considered icons of public service. Unfortunately, they've become dependent on the public dole and that muddies the waters even more. Even with their presumptive hallowed economic and emotional position, they still charge for education. I don't get it. Ha!

Meanwhile out on the farm, the trade of work-for-education is happening without any trappings of government subsidies, boards of visitors, football teams and cheerleaders. Instead, our trade is transparent and simple. The sweat dripping down the faces of these young people bears testimony to something visceral and tangible being accomplished. Any visitor or government agent can see the piles of weeds, carts full of cabbage, and baskets full of eggs. When real people do real work it's quite easy to notice and the progress is palpable.

In addition to being transparent, this relationship bears an additional burden: the prejudicial stereotype that farm work is beneath most human dignity. After all, it was indeed primarily the work of slaves in a (thankfully) bygone era. When you combine the transparency with the assumed lowliness, it's a double whammy.

How much education is necessary to justify the farming trade? In recent memory (but not historically since on-farm apprenticeships would have been considered normal in the past), it's likely that the charge of exploitation comes from farm internships that actually don't teach very much. Interestingly, several colleges around the U.S. now offer quasi-internships as part of their sustainable agriculture diplomas. These are great because they incorporate academic components like written papers and scientific data collection along with hands-on experience.

Many farm internships *are* too simple. Why? Because too many farms are simple. A farm that only produces apples, for example, will not offer gardening and livestock opportunities. A vegetable operation should also have season extension through hoop houses, floating row covers, and even a root cellar to help round out its learning opportunities. On narrow-spectrum farms, most of what is taught can be learned in the first week, and the rest of the season is rote repetition, chores rather than ongoing learning. Better internship experiences occur on diversified farms, preferably one with both plants and animals, and several species of both. The more varied the enterprise, the more there is to learn. The most fertile educational climates grow out of complex, multi-faceted farm businesses.

If the only activity is weeding produce, or picking and packing vegetables, the lion's share of the operation can be learned in a few days. The more different kinds of projects a farm undertakes, the longer it takes for the work to compensate for the education. This is why good internships intermingle brawn work with brain work. Not that we shut our brains down in brawn, but you get the idea.

The average job in an industrial processing plant, whether it's boning chicken or packing lettuce, can be taught in twenty minutes. Imagine spending your life doing something that can be learned in twenty minutes. Does that sound exciting? Any job simple enough to teach in twenty minutes uses only a few muscles over and over again. Eventually, repetitive motion disorder and carpal tunnel syndrome kick in and the worker must learn something else. No sustainable lifetime jobs exist that can be learned in twenty minutes.

George Henderson, who operated an internship program in Great Britain from around 1930 to 1960, and wrote books about successful farming, was himself an intern as a youth. Born in the city but with a yearning to be a farmer, he interned with several different operations before launching his own hugely successful farm, Oathill, with his brother. Henderson wrote his first of three books, *The Farming Ladder*, in 1944, and was truly a renegade iconic farmer during his lifetime.

Because Henderson is perhaps my favorite real farmer-writer, I want to quote him here at length. In all my books I try, at least once, to quote some lengthy passage from long ago in order to preserve these prescient pieces in our day. If people balanced modern computer habits with reading old books, we might have more reasonable applications of historic and contemporary information. Here's his take on labor from *The Farming Ladder*:

> *When we had sufficient stock to need labor, we found the only way was to have boys from school and train them as we wanted them. They were quite easy to get, as few farmers will be bothered with boys of fourteen, whom they will lose into local industry after a year or two. We did our best to teach them, and they served us quite well. We also tried to interest them in a profit-sharing system, but in their lack of general education, and their inbred suspicion that farmers exist for, and by, doing down their employees, distrusted it and preferred to have a shilling or two above the standard rate for the district, which they could spend by the next Friday night, rather than draw a substantial bonus at the end of the year.*

> *While I was learning farming I made up my mind that when I became a farmer I would take pupils, not with the object of obtaining big premiums or cheap labor, but really to teach them their trade as I would like to have been taught; I also intended to remunerate them at the real value of their labor, so they could get a better start in farming than I had had. I believe then as I do now, that the only thing wrong with British agriculture was the lack of really capable and progressive farmers, and well-trained workers. I believed too that the solution of nearly all the farmers' difficulties lay in their own brain and within the boundaries of the farms; and that far more could be achieved at home than in passing resolutions at the local meeting of the National Farmers' Union, designed to bamboozle the Government into bolstering up inefficient methods at the expense of the taxpayer. At the same time I felt that no-one is qualified to teach until he has proved his own*

theories. So we did not take a pupil until we had established ourselves as tenant farmers and then bought the farm freehold by our own efforts.

So carefully thinking out a fair system, we started with one boy. The difference was so striking between him and the local labor that for several years we have run the whole farm with pupils. Our normal staff is three. We set out to give them a really straight deal, and have been repaid a hundredfold by their loyal and wholehearted service. Most of the credit is due to my mother and sister, who keep them happy, comfortable, and well fed in the house. Outside, we teach them their job, pay them on the profit-sharing system, in which we tried to interest the local labor without success, and by which our pupils can earn sufficient in a few years to take a farm of their own, if necessary with financial assistance from us; for where could we find better investments than in backing those we have learned to know and trust, and who have thoroughly mastered our foolproof method of making money in farming?

To make sure this doesn't get lost on our current question, realize that Henderson took his pupils in the early teen years. They stayed with him for several years and their pay went up accordingly. At Polyface, once our interns have gone through their program, the ones who move on to apprenticeships receive good remuneration. Those who stay on yet another year and join staff receive more than the average salary for our area.

One of the big problems in the American culture is that we don't have a recognized middle ground between child/student and employee. You're either one or the other. Either you stay under your parents' guardianship and in school, or you become an employee. The reason people can't comprehend the internship path in farming is that we don't have a cultural norm for it today. Although many businesses, as we've already discussed, do offer apprenticeships, all of them I'm aware of require minimum wage compensation and other worker guarantees. Indeed, many if not most enjoy special government subsidies through state and federal labor departments to encourage young people entering the trades. These have all been negotiated by business lobbyists.

Unfortunately, even the farm lobby, which is huge and receives massive subsidies, has not seen the value of getting taxpayer-funded help to encourage beginning farmers in internship programs. Indeed, the mainline farm lobby is under the impression that we would have a much more efficient food production system if we had fewer farmers, more machinery, and more pharmaceuticals. In a sophisticated culture where farming is not considered normal, neither is learning how to do it, especially in a weird arrangement like internship. Anything that doesn't fit the norm is at risk of being seen as exploitation.

All of us are culturally myopic, assuming that we do things the best way simply because we don't know any other way. Perhaps a look at how things are done elsewhere would help us understand that not every people group sees things like we do. My brother Art for many years was a missionary pilot in Indonesia serving interior mission stations with airplane service. Flying supplies, medical emergencies and other things, his little Cessna could cover in twenty minutes what would take three days or more on crude jungle paths. If villagers wanted their children to get a better education, they had a system whereby young teens served as house-help in cities with middle class families.

These teens enjoyed the better educational opportunities a city provided, and broader life experience they could take back into their villages, while staying under the guardianship of their host family. My brother and his wife had several of these "house girls" during their tenure, and were responsible for their education, character, discipline, and love life. In that culture, such an arrangement was a stepping stone to either bringing knowledge and skills back to the village, or to move onto other occupations or endeavors in the non-village world.

In contemporary American culture, we have no such hybrid between home and full-time employment. As a result, life models that don't fit neatly into one of these two situations are often suspected of exploitation. That's a shame because every transition in life can't be neatly compartmentalized. And healthy, conscious, life transitions, like the house-girls situation just described, can offer smoother growth and more opportunities.

Bartering work for education *is* acceptable. The big question is, *How* and *when* does the work balance with the education?

When the two obviously balance out, then compensation is in order.

In full disclosure, I'm opposed to all governmental interference in private contracts. In other words, if two parties agree to something, the deal is between the two parties alone and it's none of the government's business. That means, of course, that I'm categorically opposed to minimum wage laws. If I want to work for you for nothing, just to gain experience, or love, or whatever, that's my right. No bureaucrat has the right to interfere with the details of free-will private contracts.

Voluntary arrangements are not coerced contracts. Coerced situations, or hostage arrangements, are a different story and fall under extortion. But voluntary agreements, made by parties of their own free will and discretion, are completely and utterly free of regulatory intervention — or should be, as guaranteed by the U.S. Constitution. Of course, the Constitution was trashed almost as soon as it was written, so it's no wonder that this recognition of basic human contractual autonomy has been abused and essentially abdicated along with the rest of that revered founding document.

Clearly, concerned observers want to know when, in the case of internships, does education stop and work start? When does education stop and work start? If I'm out in the field with an intern fixing fences and we're discussing free trade and agriculture subsidies, is the intern working or learning? This question harkens back to our earlier discussion about disassociated learning. If you're discussing fractions with your child who is helping measure out a recipe in the kitchen, is that learning or working?

When some egg-headed bureaucrat or academic tells me I'm exploiting a young person by requiring her to gut chickens while I teach her about chilling protocols for food safety, I want to scream, "Why are you pushing disassociated learning?" The fact is that if we look for teaching opportunities, almost any work situation can be done brainlessly or with the brain engaged. Your choice. Work situations are far more than simple rote activities; they facilitate learning the why behind the how. The real life working climate sets the stage for the real life informational transfer. This is such a basic intuitive axiom that to belabor the point seems tedious. But too much institutional learning, and certainly government attitudes regarding education and employment, segregate the working and learning as if they do, should, and must occur on separate planets. When an intern receives instruction in the moment, at the apex of need-to-know, it sticks.

I well remember filling out Immigration and Naturalization Service paperwork to take a foreign intern and the form asked how many hours a week would be devoted to training and how many to working? I answered with a scathing charge about uniting work and learning in all situations. They approved us. As a culture, Americans are a Greco-Roman, Western, linear, reductionist, industrialized, compartmentalized, segregated, disconnected, systematized, individualized, parts-oriented people. Specialization is one of our defining characteristics. We know more and more about less and less. Where is the eclecticism? Where is the broad-based education that incorporates practical and intellectual stimulation? The result of all this compartmentalized thinking is that classrooms are no longer a place to do meaningful work, and meaningful workplaces are no longer acceptable as classrooms.

The problem is that our culture, reflected by bureaucratic regulatory policy, doesn't recognize a two-fer arrangement. You're either a student or an employee. If you're a student, your housing must comply with university dormitory requirements. If you're an employee, your housing must comply with foreign farm worker housing requirements. If you're a student, you can't receive any compensation. If you're an employee, you can't receive uncompensated instructional time. Farmers like me who run internship programs commonly run afoul of the labor police for violating minimum wage requirements, overtime pay, worker housing standards, withholding (Social Security payments) and a host of other infractions. This regulatory climate is probably the single biggest deterrent to farm internships.

Would-be mentors don't want to spend all day filling out forms and submitting to inspections. Tragically, many would-be highly motivated and capable young people are precluded from realizing their farm dreams — and elderly farmers are forced to sell their farms — due to this workplace intimidation and extortion.

Rather than recognizing the positive value of transferring farm knowledge to the next generation, labor police criminalize voluntary agreements between consenting adults.

Too often, labor bureaucrats interpret labor law as allowing only two possibilities. If it's an educational endeavor, the student cannot accrue profitability to the organization. If the student does anything that helps the teacher make a profit, then it's deemed work rather than education, and therefore subject to all the mandates of labor regulations. At least on a farm.

If the student farmer performs meaningful work, or does anything that brings profitability to the teaching organization, then the government says it's not an educational endeavor. This myopic view separating practical learning and life experience is reprehensible to any reasonable person. But to a culture fixated on worker rights, it shows how convoluted bureaucracy can twist noble intentions.

Here I am, out in the field, moving 500 cows from one paddock to another, with two interns in tow. Prior to taking our positions, we discuss herd behavior, flight zones, lead cows, speeding up, slowing down, and turn angles. After this discussion, we take our appropriate stations and I help the interns to stand exactly where they should. Instructed to keep hands at sides and move deliberately rather than sporadically, to not clap, talk, or sing, the interns and I gently begin moving the 500,000 pounds of animals to another paddock.

Inadvertently, an intern gets too close to a cow. The cow pivots around and darts out of the herd, followed by several more. I shift to the collapsing edge. The intern, realizing her mistake, immediately backs away from the herd to take off pressure. This is a real life field study. But before the move, we used a whiteboard to discuss body position and animal dynamics based on gentle handling methods. We had theory. Now we have practice. Temple Grandin would be proud.

The twenty escaped cows gradually come around the intern, who is now standing stock still, and rejoin the herd. After the last one goes through the paddock gate, we close it and reconnoiter. The move went well — not perfect, but well. Do you know how much skill, how much finesse, it takes to move a large herd of cows like that without unsettling them? Was this exercise work, or education? My preparatory explanations and post-move debriefing — could we call that education? That I was using the two interns to help Polyface profitability is almost incidental to the job we did. How were these two interns ever supposed to test their abilities except in a real world situation?

We could talk about moving cows all day, draw a million diagrams on the white board, and give a hundred pep talks. But those interns would never become proficient if they weren't allowed to get out there among the cows and experience it themselves.

Think of the experiences people buy. From climbing Mount Everest to scuba diving on the Great Barrier Reef to African photo-safaris to big game hunts in Montana; nobody begrudges extreme sportsman and recreation-seekers their billions of dollars spent. I guarantee you that moving a big herd of cows is every bit as experientially awesome as any of those escapades. Yet at Polyface we give it away to interns for free and feed them and house them to boot!

Ever since we began our farm internship program, I've been admonished by numerous people to charge a tuition. If it's as valuable as we say, after all, shouldn't it be worth a few thousand dollars? But in my view, this would limit our candidates to those who could afford it, and I want a "whosoever will may come" approach.

What if colleges offered their education free and financed their businesses from a tithe collected post-graduation? Wouldn't that change things? Perhaps the number of courses in social justice would drop and the courses in entrepreneurism would increase. Such a change might even bring more innovation to our culture. The entrepreneurship courses might often be taught by the very people who overcame social injustice with spirit, attitude, and hard work.

If we charged tuition we would create a financial determinant in the opportunity, and we never want to do that. Rather, we offer room and board plus a stipend. The board component is out of our farm production. If they want things to eat that we don't produce, like peanut butter, that's when they use the stipend. Our objective is that this experience not be a net financial drain on any intern. We don't expect them to leave with a pocket full of money, but neither do we want it to be an economic liability.

Separating education and work as government policy-wonk elitists tend to do, denies both mentor and intern an efficient way to transfer practical knowledge. This is the great road to Hell paved by good intentions.

I hope everyone who reads these pages will become an advocate for true educational freedom. We could even tie it to vocational choice. We could even say that voluntary mentor-intern relationship rights should not be infringed. How's that for a constitutional amendment? After all, these are consensual relationships, not unlike those forged in marriage, which has been expanded to far more inclusiveness than in prior times. Isn't it amazing that people old enough to fly F-16s in combat, release drone missiles on the other side of the world, marry and have children aren't able to enter into consenting mentor relationships because the government prohibits such activities through the Fair Labor Act and the Department of Labor? Doesn't such interference smack of tyranny and patrimony?

Those who charge exploitation in farm internship situations haven't a clue as to how much it costs the mentor. To vet, house, feed, referee squabbles, and keep the labor police at bay takes an enormous amount of time and energy. In our family we've discussed many times the advantages of forgetting internships and just going with employees. We've tried to hire people who say they want to work on a farm, but too often they don't aspire to enough to be useful. This is true at least for Americans. I know this is a touchy subject so I'll let it drop there.

Henderson found a similar situation. In his books he explains how he tried and tried to find good farm workers, paying well above going wages. But at the end of the day, they were just farm laborers, not interested in betterment. When he began his apprenticeship program, however, he found young people ready to do great things in life and the quality of his farm staff improved greatly.

If Polyface solely had employees, we wouldn't have to deal with the turnover and the orientation of new recruits. We do have an extremely loyal, faithful, and capable core of employees and subcontractors on our staff. But the interns radiating out from that center keep us from becoming insular. We believe that Polyface prospers as we enable these young people to prosper. One of the advantages of interns is that if we don't like one, at least he'll be gone in a couple of months.

Having interns is far riskier than having employees. Employees fit neatly into the proscribed box handed out by the government labor extortionists. Interns, because they don't fit neatly into that box, subject the mentor to a bureaucratic gray area, the murky waters of investigations, audits, and possible arrest. It happens.

Because interning is marginalized culturally and governmentally, interns need to realize, with deep gratitude, the risk and inventiveness undertaken by mentors. I don't know a single farm mentor in America who's comfortable in the role. I don't mean comfortable teaching and enjoying young people. I mean comfortable in wondering if a letter will arrive from a government labor extortionist demanding square answers to round questions. Demanding segmented explanations for integrated procedures. Demanding compartmentalized descriptions for immersion-type knowledge transfer.

As one who's been clandestinely practicing in this gray area for two decades, I assure you that these concerns are never far from my thoughts. Yes, we try to stay legal, and I'll talk about that in other chapters. But after doing everything possible to comply with the letter of the law, I guarantee you that any labor extortionist bent on criminalizing will find some i not dotted or t not crossed. Good grief, the federal register is so voluminous now that anyone can be arrested for just about anything.

It's outrageous that the U.S. now has nearly twice as many people incarcerated — roughly to the tune of $50,000 a year per prisoner — in America's prisons than we have farmers. At least the prison population is big enough to merit demographic notation on Census Bureau charts. Not farmers. This should spawn a revolution.

For those of you having trouble seeing the connection, let me make it extremely clear. Whenever a culture asks for more centralized solutions to problems, everything becomes more centralized. In other words, if we ask for more federal government involvement in anything, and I mean anything, that thing, whatever it is, will gravitate toward undemocratic centralization. The more we ask for government intervention in agriculture, of any amount and any type, it will create a more centralized, undemocratic agriculture. That centralization manifests itself in many ways:

- Harder for fluidity between generations
- Harder for small farms to survive
- Opaqueness in the marketplace
- Bigger retail venues
- More paperwork to comply with government rules
- Expanding factory farms
- Bigger equipment
- Neighbor-offending farming practices like odors, pollution, or pesticide drift
- No trespassing signs on farm gates
- Laws barring photography at processing plants
- Mandatory radio-frequency animal identification
- High insurance premiums for direct-marketed products
- Difficulty becoming a vendor for large food distributors

Rather than being accused of exploiting young people, farm mentors who do farm internship programs right should be placed on a pedestal as icons of righteous virtue. Unless and until anyone has a better solution for transferring information imbedded in the agrarian subculture to our best and brightest young people, I submit that farm internship programs represent one of the most powerful and direct beacons of hope for creation stewardship.

Do we really want farms turned into more golf courses? Do we really want farms gobbled up in an orgy of bigger farms? Do we really want farms abandoned to wild lands and wastelands, offering no employment, no stewardship, no tax base? I submit that vibrant rural economies are far more important than our current urbanized and food-disconnected culture realizes. Today's interns are tomorrow's farmers, the foundation for vibrant rural economies.

I'll love them, care for them, encourage them, teach them, springboard them wherever they want to go. Every country in the world would be better if its fields were stewarded by the kind of young people who come through our intern program. They are the most enjoyable, fun-loving, hard-working, conscientious young people you could ever imagine.

We have thousands of visitors a year come to Polyface. In fact, we have a 24/7/365 open-door policy for anyone to come from anywhere in the world to see anything anytime unannounced — that's our commitment to transparency. From high society couples to hard-scrabble business owners, visitors routinely remark about our happy interns. Did you catch that? — happy. H-A-P-P-Y. Smiles are the first crop we cultivate. The demeanor on these young people indicates something far different than exploitation.

Any intern can leave at any time. When we began this program, Teresa said what she liked about it was being able to have more children without becoming pregnant. That's exactly the way we see it. They apply and we pick them. They pick us. And then the magic begins.

Chapter 4

Mindset

A re you ready for interns? This isn't a question to be taken lightly. The day you start taking interns is the day your own daily life will be subject to the microscopic scrutiny of others. From your beliefs to your actions and even the language you use, interns will probe your every thought and action. The interns don't mean to badger or argue; in good faith, they want to know what makes you tick.

Interns analyze every single thing you do. If you have more than one intern, they'll discuss your words, your church, your family and come to certain conclusions. Because they're coming into your life without benefit of history or perspective, they often make incorrect assumptions and draw mistaken conclusions. It can be maddening at times. You'll have to choose which perceptions are erroneous and important enough to debate and the ones that aren't worth a fuss.

Be assured that interns will scrutinize, analyze, and pass judgment on every aspect of your life. They're kind of like children that way. Sometimes we parents need to pull our children aside and explain crazy Aunt Matilda, and why we interact with her the way we do. Other times, we're not even aware of how we're dealing with someone--the odd phrases or body language because it's become habitual, but the children fill in the blanks over time. With interns, all of these familial nuances occur within a few months. That's why I ask for lots of tolerance from interns. It's also why mentors need to be patient and tolerant when interns form their opinions.

To be successful, mentors must approach the whole program with the right mindset. For the best outcomes, both parties must come to the table with right perceptions, and that is partly what this book is all about.

The expectations and objectives on the part of the mentor shape everything. The difference between starting an internship program to procure what you think will be cheap labor, or starting one in order to send off successful practitioners to other farms is pronounced.

No mentor should start a formal program without answering this question: Why am I doing this?

If you're doing it for so-called cheap labor, don't ever start. You'll find out very quickly that interns are not cheap labor. Their constant mistakes, lost tools, and broken machinery will have you wondering about the sanity of anyone starting an intern program. You'll probably lose your own mind.

A much more lofty goal is necessary to sustain you through the frustrations of having interns. Perhaps the biggest surprise for mentors, who expect to whip young people into shape, is that this learning experience is a two-way street. In reality, Polyface interns have had as big an effect on me as I've had on them. As I look back over our twenty years of internships, I confess a dramatic shift in attitude.

Early Lessons in Internships

Polyface was rocking along well and our two children were old enough to be a huge help. As a successful working farm family, we were fulfilled, happy, and satisfied. When we became more successful and I began conducting some seminars on our methods, young people began contacting us about apprenticeships. We had a neighbor and some friends who came over to help us when we needed it. Of course, we helped them when they needed some extra help, too. Farm labor swapping has always been a foundation of the agrarian community.

Although we took the apprentice requests as a compliment, we didn't feel like we had enough work to keep an apprentice busy. Besides, where would he live? We certainly didn't want our private home complicated with a live-in of any persuasion. For us, the program developed serendipitously.

We had always maintained an open door policy for people wanting to come and learn. Routinely folks would come and process chickens with us, for example. Working in such proximity, we could teach them but also keep a sharp eye on their technique without leaving our work stations. It was a tight work area, yet we welcomed people to come and learn.

One summer a customer family asked us if they could bring their son over to have a farm learning experience. He was only fourteen and Daniel was thirteen, but they were friends as well as customers and shared our moral values, so we said yes. Every weekday morning for nearly two months Sam's mother would drop him off with his lunchbox and he spent the day working with us. Because he was near Daniel's age, their compatibility created a friendship that made the relationship better. Being so young, it was easy for Teresa and I to view him as a child — a son like our own.

He and Daniel became fast friends and Rachel enjoyed having another brother around. This initial introduction went fairly smoothly and piqued our interest regarding internships. This initial experience helped me to understand how much I enjoyed seeing someone else learn. With Daniel, it was more like osmosis--it developed slow enough that I didn't seem to notice. But with Sam, going from not even knowing what a digging iron was to wielding it effectively as a post tamper was dramatic. Totally unaware of my own deep fulfillment in mentoring someone into my skill set, this summer helped me discover something about myself I didn't even know: I enjoyed helping young people grasp new skills.

A year later, a delightful young man, Zach, who had just graduated from the University of Colorado, contacted us, desperate to work at Polyface. He came for a visit and threw himself into the work with fervor, but we weren't ready for a real full-time apprentice. Showing great creativity and persistence, he found a job as a videographer on a whitewater rafting outfit in West Virginia.

He kayaked ahead of the raft, filming it going over rapids, then sold the video to the rafters at the end of the run.

Since these excursions only ran on weekends, he worked out a schedule that enabled him to be at the farm Monday-Thursday. Friday morning he drove out to West Virginia, spent the weekend as a kayak-videographer, then came back to the farm either Sunday night or Monday morning. Not having any separate housing, he spent his nights up in the attic of our house — where Teresa and I spent the first seven years of our marriage. In 1987, once our children came along, Dad and Mom had put in a house fifty feet away from ours so Teresa and I could have more room in the big old farmhouse.

Zach was a delightful fellow who added new perspectives to the farm. He stayed through the summer and then headed off to South America to run some more rapids with his kayak. We were dabbling in opening up the farm to these kinds of relationships and found having another young person around fairly satisfactory. During this time, we realized we didn't want to house interns in our own house. It was just too invasive. The American culture wants some space and farmers are no exception.

Shortly after Zach, we had a customer with a son who was struggling. Her request was to take him in as a kind of informal intern to help him get straightened out. Apparently he'd had some drug issues at college, bombed out, was directionless and drifting. In deference to this nice customer, we agreed to take him as long as he didn't stay with us. His mother brought him out every day.

That experience was an epiphany. To protect his identity, I'll call him Jasper. We discovered quickly that we didn't want to be a reform school. The quickest way to be turned down for a Polyface internship is to have your mother write us a letter. If the would-be intern isn't personally and individually initiating, pushing, persisting, and negotiating for the experience, forget it. Certainly plenty of farm-based reform programs or juvenile delinquency programs exist, and they're wonderful. Some do outdoor camping and adventures, some homesteading and draft animal skills, others go on extended wagon train travels--in all cases, interacting with nature, animals, responsibility does far more than institutional incarceration. But that's not us. We're just an ordinary working for-profit family farm with enough weirdness of our own. Ha!

That's why you have to understand clearly what your goals are. Your expectations define how much breakage, ignorance, and laziness you can stand. If you're running a reform program, make sure the farm is not your livelihood. For us, having Jasper was a disaster. But we learned from it. We learned who we are and what we want our farm program to be. That much was helpful. Coming on the heels of two positive experiences, it showed us the gravity, the seriousness, of bringing interns onto a farm. When we hear Jasper's name, twenty years later, we still shudder with horror. Be assured that bad help is much worse than no help at all. Jasper couldn't follow any instructions. He was lazier than a sloth. The final straw came one afternoon when he used the drill and broke not one, but two 3/8 inch bits in a matter of minutes.

Now dear folks, I've been drilling things for a half century. Hard oak boards, locust posts so hard the sparks fly, and countless pieces of steel in farm construction projects. Upside down, right side up and sideways. I've rigged up my fair share of poor-boy drill presses using chains and a pry stick. And I have never, ever, ever broken a 3/8 inch drill bit. That's practically half an inch! Anyone who has done much drilling has broken tiny bits, like 1/16 inch or smaller. They're extremely fragile and any novice will break a few on the way to gaining skill.

But a 3/8 inch bit? That's practically as fat as your finger. In fact, if you were to ask me right now how to break one, I'm not sure I could tell you. I suppose you could put it in a vise and smack it with a heavy hammer, but it would still take quite a swat. How do you break that hefty a bit in a piece of wood? Do you know what's amazing? Jasper didn't know either. He had no clue. It just happened. We got rid of him that day and celebrated our escape from the curse of Jasper. Whew! We barely survived that ordeal.

Of Eggs and Ideas

The next year, in 1991, a customer asked us to raise a hundred pullets for her. Pullets are female chickens who have not yet begun to lay eggs, just as a heifer calf is a female bovine and a filly is a female horse that has not yet foaled. When she foals, she becomes a mare and a bovine who calves becomes a cow. Enough. So this customer wanted to start a small egg laying operation but wasn't equipped to start the little chicks and raise them for twenty weeks until they started to lay.

We agreed to start them for her. I called her at twenty weeks and told her to come and pick up her chickens. To my complete astonishment, she changed her mind and didn't want them after all. Suddenly we were stuck with one hundred pullets. Interesting thing about those chicken women (my vernacular): when they begin laying, you can't ask them to stop. "Can't you put a cork in it?" doesn't elicit a positive response. So the eggs started rolling in.

We hadn't planned on such an influx of eggs and it overwhelmed our market. We began offering "Buy 10 Dozen Get One Dozen Free" deals to our customers. One of our long-standing customers came down to get his broiler chickens (dressed meat birds) during this time and was appalled at our egg problem.

"This is immoral!" he cried. "These are the best eggs in the world, and you're having trouble selling them? That's like not being able to sell maple syrup. It's outrageous. I'll take thirty dozen."

Elated, we sold him the thirty dozen, which is a whole case. Two days later he called us asking, "May I come tomorrow and get two more cases?" Of course, we were excited to sell sixty more dozen.

He came and got those, then three days later called back asking for three cases. By this time, we knew he was up to something. He and his wife were empty nesters living in northern Virginia and I was fairly confident they were not eating thirty dozen eggs a day. I asked him what in the world he was doing with all those eggs.

Turns out he was a lobbyist in Washington D.C. for a national health organization and routinely took legislative aides and policy wonks to swanky restaurants. Being an inveterate entrepreneur and Polyface cheerleader, he'd go back to the kitchen and give some of our egg samples to the chef. Invariably, the chefs went crazy over these eggs and asked if they could buy them.

Within two weeks of his initial volume purchase, he came to us with a formal offer. "If you'll raise the eggs, I'll sell them. How soon can you get me six hundred dozen a week?"

I like eggs, but I was busy and not ready to take on a lot

more responsibility. The farm was highly profitable and we were completely satisfied with our lives. But I've always tried to meet market demand, even if I didn't know how at the time. It's not because I'm interested in growing a business empire, but because it's a way to get healing food into more mouths.

I told him that it would take us a month to get the chicks, five months to grow them up to laying age, then another month for the eggs to come up to saleable size. The first eggs a chicken lays, called pullet eggs, are quite small. It takes a couple of weeks for the size to increase. A total of seven months. He agreed that was fine.

Now we had a huge decision to make. Do we proceed? If so, how? Since the trouble with Jasper we'd been turning down apprentice requests. At the same time, we still had the good memories of more successful interns. Suddenly I had an idea. What if we created a separate business within Polyface and delegated it to an apprentice? Then the apprentice would have personal ownership while we shared the risk and profits. Perhaps it's important here, for clarity and story flow, to explain that in the early years, we did not have interns, which we see as shorter-term young people. We started with year-long apprenticeships and only after we'd been doing that for ten years did we add a four-month intern program. Anyone we've had for a shorter period of time than a full year we tend to think of as an intern. Now back to the story.

I saw this 600-dozen-a-week egg enterprise as an opportunity to finance housing capitalization along with a way to say yes to these wanna-be apprentices who were contacting us. Polyface had a substantial vertical growth step that year. It was similar to the growth that occurred right after Michael Pollan featured us in his bestseller *Omnivore's Dilemma*. Businesses never change in a nice straight line--their changes are always a series of stair steps, and this one was a tall step. We ordered a thousand chicks and set a ball in motion that would fundamentally change Polyface starting a formal apprenticeship program.

We purchased an outdoor storage building from a local builder who constructs them on site. Using their 12 feet x 20 feet carriage house option, we situated it about two-hundred yards from our house. Because our farm lies in an agricultural zone, which prohibits other dwellings, we poured four concrete pillars and attached two 20-foot I-beams on them. We bought four 10-foot pieces at the scrap metal yard and I butt welded them to get the full length. Then we affixed the floor joists of the cottage to the I-beams. By doing that, we took the position that it was a portable structure and therefore a farm machine. We could easily slip a hay wagon under it, unbolt the I-beams, and move it somewhere else. By calling it a farm machine it technically was not a building, not subject to inspection or license, and it did not go on the land as real estate. Yes, it definitely pushed the envelope, but remember, I wrote a book about *Everything I Want To Do Is Illegal*. We ran power and water to it and welcomed our first apprentice right after the pullets began laying.

But we had problems.

Our entrepreneur lobbyist got busy jet setting around the country dealing with municipalities that wanted to add fluoride to their water. Suddenly we had a thousand hens laying eggs, not just a hundred. These chicken women were no easier to control than the previous chicken women. How we solved that will be a story in some future book about marketing, but this sudden wealth of eggs is how and why we entered the apprentice world. It was never to get cheap labor. Rather, it was a partnership to train up someone who would "go thou and do likewise."

From day one, our primary objective has been to pass on farm knowledge, equip our interns and send them out to be farmers in their own right. In my view, that is the only motivation that will sustain a mentor through the ups and downs. A close permutation, of course, would be to train up a farmer who could take over your farm.

And Apprenticeships Were Born

Tai, our first formal apprentice, was a quick study. After the disaster with Jasper, Tai was a breath of fresh air. We liked his family, his spirit, and his enthusiasm. He moved into the brand new apprentice cottage, complete with crude composting trash can toilet, and we started down the formal apprenticeship path.

Geronimo then came from Mexico and our cottage, built for one, received a bunk bed and housed two. They got along famously and life was generally good. Except for one thing: tools began disappearing.

This is where mentoring begins shaping the mentor. Ironically, with more hands on the farm — and I mean this literally as hands, not farmhands, but just hands — the ability to hold things together gets much harder. The more people you have, the more different ideas, temperaments, and objectives you have. The whole outfit can unwind into a million loose threads. That is the maddening part.

People who know me now can't imagine that I used to lose my temper quite often. I know this isn't unique to me. But I'm confessing the problem because it speaks directly to the mindset of the mentor. My words vacillated between complimenting and berating. That's a big emotional range, a yo-yo demeanor. In short, you can't do that.

Steven Covey in his *7 Habits of Highly Effective People* talks about emotional equity like a gas tank. You have to put enough in to make withdrawals. All of us make withdrawals from time to time. But if I may stay with his analogy, he's assuming a measured withdrawal, not bashing in the tank with a pickaxe. Get the picture? There's a big difference between an unfit word and a waterfall of excoriating invective. I never cursed, but it was just as bad.

I would just lose it, becoming so frustrated with the ignorance of interns that I would vent — verbally exploding that emotional gas tank with my inflammatory words. No amount of emotional equity can sustain that, even if the occurrences are spaced widely. Finally, Jason took me aside in the early days of his apprenticeship and held me accountable. After a particularly bad episode--right before I headed off on a speaking engagement to tell people about our wonderful farm--I apologized to him. I do cool down pretty fast, and have always had a gift of forgiving and forgetting. Not wanting this awkward incident hanging over my head while I was gone, I wanted to make it right and get on with things. "Jason, you know this isn't right and it's not me. You know it doesn't happen often, and I'm sorry."

Leveling his eyes, he said he forgave me, but I was wrong: "It happens far too often." Wow. That hit me like a ton of bricks. Showing the maturity of a man beyond his years, he held me accountable. While I was away for those couple of days, he penned one of the finest examples of charitable criticism you could ever imagine, full of grace and appropriate gratitude and encouragement, but balanced with the honesty that the situation required. In conclusion, he asked for an audience when I got home. He got one. I cried. I had a reckoning with God that weekend, and I've never yelled at an apprentice since. The apprentice, in this case, became the mentor, and it was one of the most precious experiences of my life. He continues to be a shining example of thoughtfulness and openness, a true friend for life.

The mindset a mentor needs is essentially seeing farm interns as kindergarteners. I know, I know, they look like adults. They generally come potty trained and able to handle fork and spoon at the table — sometimes. But their awareness of all things farming is zero. We noticed this immediately because Daniel, who was only fourteen or so when we started with Tai, was light years more aware of farming than these twenty-year-old apprentices. He knew whether the tractor was going forward or backward. I kid you not. This is basic stuff.

Daniel knew that when I was cutting a tree with the chainsaw, it's a good thing to keep your eye on the tree to see where it's falling. He knew which piece of wood I was going to cut next and would toss it up on another log so I could cut the first one safely and comfortably from above the ground. You can't cut a piece of wood with a chainsaw if it's on the ground because the saw tip can easily dip into the soil and hit a rock. Chainsaws and rocks are not friends, trust me.

This is why we don't allow interns and apprentices to run my chainsaw. We have a communal saw that everyone else can run. But I have two untouchables: my chainsaw and my wife. Ha! Daniel's situational awareness is almost like telepathy, but these green interns are as unseeing as a fence post.

As a mentor, the only way I know to compensate for a young person's ignorance and inexperience is to view them as your own children when they were toddlers. You don't fuss at your own children for their clumsiness as they begin to walk. If they fail to obey, you don't lose your temper and berate them. You discipline them, and that's another topic that I won't address here, but you don't lose your temper.

For sure, especially for men, losing your temper is probably a much bigger problem in our society than most realize. Although I was always able to hold it with my family; I found it combustible with interns. I don't know why, except that when someone looks like an adult, you just assume they know more. Certainly they know about computers and Facebook and video games. But using a crowbar or separating cows in a corral? That's a different story.

I'm being as transparent and personal here as I know how to be. After Jason checkmated me, I became a much better person. Frankly, as a result of interns, I'm a much more mellow, forgiving, charitable, patient person. Not perfect — oh no, not by a long shot. But I don't think I ever would have grown spiritually and emotionally to this point without apprentices. It's heartfelt when Teresa and I say we love interns like our own family. Interns have brought tremendous joy and growth into our lives.

If you assume that interns don't know anything, you'll find yourself slowing down in your instructions and trying to create learning opportunities in every work situation. This relationship is entirely different than employer and employee. Thinking like an employer will make you crazy because you'll be so disappointed and frustrated. Don't even think of it as boss and underling. Think of it as parent and child. That's the only way to have a healthy attitude that creates a climate helpful for the intern.

Think of how you teach your child to wash dishes. You don't just all of a sudden one day announce to your four-year-old that, "Hey, I'm going to town to run some errands. Wash these dishes while I'm gone and I'll see you in a couple of hours." That's ludicrous! What we do is invite the youngster to stand on a stool next to us and wash with us. Doing it together, over and over and over and over, is what transfers our ethics, standards, and attitude to the next generation.

The old adage "more is caught than taught" is certainly true. The mechanics of how you mentor pales in comparison to your mindset — your spirit and attitude behind the mechanics. We've all experienced instructions given in a way that make us eager to help versus the ones given with a barb. A sarcastic edge on the instructions nullifies all the mentoring you're hoping to offer.

Note the difference between two examples: "Could you take the four-wheeler up and check the pigs? To make sure we're on the same page, that means spark, water, wire, and feed."

Here's the same thing with a barb: "Do you think you're finally capable of taking the four-wheeler and checking the pigs without your hand being held? By the way, that means spark, water, wire, and feed. " The differences are subtle, aren't they? But the spirit of the instruction is as different as night and day. In either case, the intern can express reservations about feeling competent to do the job, but the latter demeans regardless.

I call these instructions with a barb. In the section on interns, I'll address the flip side when I talk about questions with a barb.

Now, lest anyone interpret this as a demand that mentors be sinless saints, I assure you that occasionally I still shake my head in the presence of interns, or express my disappointment, even sternly, if gross negligence is obvious. But no matter how egregious the infraction, it doesn't justify screaming and yelling. Emotionalism like that doesn't work with your own kids, won't work with interns, and only makes you look like you're out-of-control. All losing your temper does is alienate. I'm hoping that by using the parent-child metaphor most people will understand my point that taking on interns requires being patient.

Delegation is the next key to mindset. Farmers are an accursedly independent lot. We spend a lot of time out by ourselves dealing with things, thinking through things, and getting things done. Most of us think we're doing it all the best way possible. After all we've done these things for a long time. We've put in our ten years and 10,000 hours, which are supposedly the prerequisites for proficiency.

But the whole idea of mentoring is to transfer the proficiency, for interns to learn by doing. The intern can't learn if we don't let her do the work. Our whole goal is to be able to confidently go to town and leave the child washing dishes, knowing that the job will be done efficiently and to our standards. How do we do that without delegating? How can the child or intern demonstrate proficiency without the responsibility of accomplishing the task?

"I can do it faster myself," only works for awhile. All of us age eventually and need people with proficiency to take over the

63

jobs we can't do any more. All of us will pass that way. Anyone who thinks refusing to delegate will make his life easier is not thinking long term. While mentoring may be tedious at the outset, the outcome is the freedom to move onto other things, whether it's the next life or new adventures in this one. As difficult as all this is, it's still much better than hitting seventy . . . alone.

One of the greatest blessings of my life is that I can leave the farm for a month and know that Daniel and the Polyface team will operate it exactly like I would if I were there. Do you know how freeing that is? That most farmers grow old as single-player curmudgeons is truly the bane of agriculture. Middle-aged farmers must open their hearts to young people in order to emancipate themselves from being handcuffed to the farm in older years, or at risk of losing it due to aging. At sixty years old, you can't run the farm you did at thirty. It's just not possible.

As you begin to delegate, you'll definitely have to come along with some corrections. Sometimes several times. Think about any skill you now possess and your first fledgling attempts to master it. Sharpening a chainsaw. Stacking a wagon load of hay. Shoveling. Now there's a skill. Do you know how many twenty-year-olds have no clue how to handle a shovel? Every time we do something with a shovel I have to start at the beginning. "Here's where you place your hands" — much like you show your kids how to place their hands on a baseball bat.

It seems almost silly to show a strapping twenty-year-old how to place his hands on a shovel. But it's necessary to keep from being anguished at the painful inefficiency exhibited by novice use. Sometimes I have to just look away because it's too painful to watch. Interns throw across their bodies, don't take advantage of leverage with their hands, and seldom know how to use their whole body to proper effect. I can't tell you how to do this. You won't read it in a book. I have to show you. Never lift the shovel without a full load. You can jiggle the shovel, pushing it forward with your thigh, and make it work incredibly efficiently.

As we work together, I have to delegate the shovel work at some point to make sure the intern gets the instruction. The goal is for proper shovel use to become routine so that I can leave a job and know it's being carried on expertly. The idea is to duplicate myself. My dad always said his goal in any job was to make himself unnecessary.

As a tax accountant, Dad's greatest joy was in teaching a youngster in a family business (usually a daughter) how to keep the books. Routinely this resulted in fewer billable client hours for him (isn't that a novel idea?) but it allowed him to in turn show another family business how to keep the books. His goal was not billable hours per client, but independent clients. He essentially delegated the rudimentary parts of his work in order to spend more time on the fun parts, like setting up meaningful books for small businesses or figuring out clever loopholes to save taxes.

As a farmer, if you had unlimited time and unlimited money, what would you do? Think about it. To Wall Street types, let me say that most farmers would *not* answer: "Retire and go sip martinis on Martha's Vineyard." I know that may be hard to believe, but it's true. In this I am sure: Farmers for the most part love their farms and love growing things, whether it be plants or animals. We have a universal love for being viscerally involved with landscape husbandry.

But within that context, we also have parts that really float our boats. Every vocation has pieces that people find more enjoyable than others. Once you figure that out for yourself, then imagine actually being there.

I'm not asking you to do something I haven't done. So let me tell you what floats my boat. My favorite parts of farming — the stuff I would do without being paid for it — are designing redemptive capacity into the landscape. This means freeing healthy trees from diseased, crooked, dying neighbors. Building ponds to hydrate the landscape. Installing pasture-based livestock systems on more acreage. Getting nutrient-dense, soil-building food to more people.

How can I spend more hours on that? The answer is simple. By freeing myself up from the more rudimentary elements of our farm, I can devote more attention to these other things, to my passions. Nobody can do it all. We aren't designed that way. We shouldn't even try. It's bad practice.

Delegating parts of the work load to partners is the path to freedom and full self-expression. Yes, I'm as good a post hole digger as you'll ever find, but that's not all I am. If that's the ultimate expression of my Joelness, it would be a shame. As mentors, we must take satisfaction from working ourselves *out* of jobs. Striving for that achievement needs to drive our mindset.

Ultimately, the mentor's mindset is one of servitude. We're serving others in order to enable them to be where they need to be. That's the pinnacle of leadership, after all. Leadership is all about serving people so well that they want to follow. That means we have to be the first ones up in the morning and the last ones to bed. We have to do the dirty jobs. On our farm, we make a big deal about the sacredness of every task.

In the final analysis, we don't have jobs of honor and jobs of dishonor. All tasks are meaningful tasks that must be done. Period. No prima-donnas. That's a sure way for an intern to get sent packing. By the same token, interns can spot a finicky mindset in a mentor a mile away. As mentors, serving others is our highest calling. If we can instill that mindset in our interns, they'll be successful no matter where they go.

Early on, Teresa and I decided to invest in our children with nice salaries, not allowances. Nobody should get paid for breathing. But as they got older, we paid them salaries. They each had their own businesses as well. Each, at twenty years old, had $20,000 in the bank, having earned every penny.

While Teresa and I could have used them as cheap labor in the later teen years, and put that money in our own bank accounts, we decided to serve our children with it. That was all part of a grand scheme to invest in them early on so that when we get old and feeble and need servants to help us through the day, our servant's attitude will grant us love and care from our children.

An attitude of service seems simple enough, yet most people don't ever implement that message. Serving each other is the highest form of leadership. For the record, lest anyone familiar with my libertarian tendencies take this as justification for raising taxes to help people through government assistance, that is not servitude. Jesus never admonished people to use tax money to help anybody. He admonished us each personally to go, to give, and to do.

To equate government violence (try not paying your taxes and see who gets violent) with putting money in the offering plate is just plain idiocy. When people invoke Biblical mandates to help the poor to justify government welfare programs, it equates individual charity with government charity. To equate the Internal Revenue Service with the Sunday morning offering plate is ludicrous in the extreme. Of our own volition, however, self-sacrificing servitude is noble and arguably one of the major differences between humans and animals. I've never seen a pig miss a meal to make sure a runt had a chance at the trough. Ain't gonna happen. Individual charity is one of the foundational ways we express our human-ness.

By adopting a servant's mindset, then, mentors can keep their sanity through the process. Lost tools, broken gates, cows out, thirsty chickens — the list of frustrations brought on by interns is endless. But a servant leader goes on, refuses to let this dampen his purpose, and continues to make a path that others can follow.

Chapter 5

Skills

I'm almost embarrassed to write this chapter. It seems like such an obvious foregone conclusion that I wouldn't have to say it, but say it I must: mentors need to have attained proficient skills at their trade.

Too often, I hear novice farmers, grossly inexperienced and wallowing in self-doubt, with inept practices obvious everywhere on the farm, almost cavalierly say, "What I really need are some interns to get this place in shape." I have news for you. If you can't run your farm, interns won't run it better. You need to be running a tight ship before thinking about developing an internship program.

My standard measurement for self-evaluation is that if I come to your farm and you have to apologize more than one time for something, chances are it's not a tight ship. Everything can't be perfect all the time, and it never will be. But it also can't look like a bomb just went off. Or that everyone slept in until noon. Or that everyone walked around with their eyes closed all day, unable to see the obvious.

Craftsmanship certainly has a subjective element, kind of like beauty is in the eye of the beholder. But at its most fundamental level, the work must possess a beauty and order that sets it apart from the competition. Farming craftsmanship is no different. This chapter is as much about knowing when you're ready for interns as it is about interns knowing whom to choose as mentors.

To my knowledge, no accrediting organization exists to certify farms as credible or not. I don't think farm-mentors have passed any sort of minimal competency standards. Farm-mentor/intern curricula have never been codified. By pointing out that this field has a bit of a Wild West flavor to it doesn't mean I wish it were codified or credentialed or licensed. Quite the contrary. If there's one true blue buyer-beware marketplace in America, this is it. And I think that's a good thing. It pushes both sides to do some research.

As a mentor, then, how do you know you're ready? As an intern, how do you know a certain farm can really teach you what you need to know? Here's a checklist that I hope will be helpful.

1. *Order.* Are the premises ordered? How much junk is sitting around? Successful farms, including Polyface, aren't golf courses or equestrian estates. They often exhibit a threadbare look. More about that in a minute. But random yards of weed-covered car bodies, fencing wire, and axles indicate neglect and shoddy practice.

Such shoddiness will be apparent in the field. At our farm, ongoing projects, in various states of completion, require a certain amount of stockpiled materials and works-in-progress. But it shouldn't assault our visual senses. I always tell my team that, "It's okay to have stuff around, just make sure it looks placed and not thrown."

Placing materials on a pallet, in a straight line, can show at least some appreciation for order. Hanging things up on a wall instead of throwing them in a pile might not change the amount of stuff at all, but it sure changes visitors' impressions. People expect a working farm to have inventory stashed on site. But it should look like it was placed with purpose, not haphazardly.

2. *Utility.* This is the opposite of glitzy. Profitable, successful, working farms have a functional look that's different than ones I characterize as gentrified. Outbuildings that look humanly habitable, lots of perfect white wooden fences, manicured lawns and spit-polished, new-looking tractors indicate that money is flowing from off-farm sources. I'm not opposed to that, but farms floating

on off-farm income are not generally the kind to teach aspiring young people how to make a living farming.

Now, if your goal is to intern for a horse farm that routinely competes in the Kentucky Derby, disregard everything I've said here. But while I encourage internships in every kind of farming — indeed, in most vocations — these showplace farms generally don't present the bootstrap opportunities that form the backbone of agriculture.

Farmers who make a living farming practice frugality. They usually have a couple of older tractors in the equipment shed. The wooden fence — if there is one — normally has a board or two loose and several needing paint. Really innovative farms might have a portable chicken shelter in the back yard, or mow the lawn with sheep. The farmer's attire shouldn't look too crisp, like it just came from a stylist's shop.

Visitors to Polyface are often taken aback when they find all of us on the team wearing monogrammed shirts with others people's names on them. We get them cheap down at the uniform cleaners. When people die on the job or get fired or quit, their monogrammed shirt goes on a for-sale stack at 50 cents per. That's as cheap as rags.

The other day I was walking between the house and the equipment shed when a smartly dressed couple visiting the farm walked past (we have 24/7/365 open door policy for people to come from anywhere in the world anytime to see anything unannounced--it's our commitment to openness and transparency). We have thousands of visitors annually and can't stop and talk to each one, much as we'd like to. I nodded a courteous hello and they asked: "Do you work here?"

"Yes," I replied.

"This is an awesome farm," they gushed. I nodded in agreement and walked on, realizing I was wearing a shirt monogrammed: Carlos. The business logo imprint was equally off-Polyface: "Julio's Landscaping." The couple had no idea who I was, and I certainly didn't lead them on to think I was anyone other than Carlos the landscaper. Sometimes being incognito is fun. I watched the couple disappear around the barn on their way out to the cows and then had a good belly laugh.

3. *Health*. Do things look healthy? I always tell visitors to look at the fields and animals, not the buildings and machinery. If it's a livestock farm, do the animals look robust and slick, or do they look tired and a bit ragged? If it's a produce operation, do the plants look verdant and vibrant, or are leaves ragged and spindly? Are the plants homogeneous, standing like soldiers, in a fairly consistent physiology, or are they highly variable, with big ones and small ones sharing the same row or bed? These visual cues speak volumes about basic care and health on the farm.

What about mud and noxious odors? A little goes a long way. Every farm has tough times once in awhile. A flood or heavy snows can create unsightliness for a couple of days, but it shouldn't be a constant feature. Good management necessitates giving animals places to lounge that aren't muddy — even if they must go through some mud getting there. It's one thing to walk through some mud; it's quite another to have to sleep in it.

I don't find a sick pen objectionable. That shows the farmer knows about a problem and is taking action. Of course, a sick pen should contain a tiny, tiny portion of the animals. If all of them look like candidates for the sick pen, probably some crisis is happening. Every farmer deals with sickness and disease from time to time, but the farm should still look healthy overall.

If the farm has laying chickens, what do the nest boxes look like? Are they pristine, and the eggs in them nice and clean? Or are they soiled with manure and broken egg yolks, making the eggs dirty and unsightly? If it's a vegetable operation, what about weeds? These are the telltale signs of professionalism.

4. *Philosophy*. Does the farm exude a clear, consistent philosophy? Mentors need to know more than just how to farm. They need to know why they farm. They need to be able to explain the deepest reasons behind what they do. Unfortunately, too many of us are just technicians and haven't stopped to really address the deep-seated why.

I have the greatest respect and admiration for two farmers who couldn't be more different in their approach. One is Eliot Coleman, guru of extended-season production in Maine, who arguably received the mantle of leadership offered by Scott and

Helen Nearing, icons of the early back-to-the-land movement. Although it wasn't formal, Coleman did a quasi-internship with them and impressed them more than any other aspiring farmer. The author of several books, Coleman's operation is quintessentially ordered. The plants and rows are completely weed free, bug free, and look as visually perfect as an architect could create. The plants look like they've been planted along a glorified ruler. They aren't just straight; they're perfect.

On the opposite end of the spectrum is Bob Canard, farmer extraordinaire to food maven Alice Waters' Chez Panisse restaurant in Berkeley, California. I'll never forget the first time I listened to Canard give a presentation at a sustainable farming conference. I was mesmerized by his talk, and then some! Opposed to green manuring, he said in order to really feed the soil, plants need to reach physiological maturity, full phenotypical expression, and turn brown. Wow, here was a different kind of guy. He hunts wildlife manure to create fertilizer tea concoctions and his vegetable gardens look like a hodgepodge of thrown together variety. His gardens are a cacophony of apparent disorder, but the taste, texture, and production are as good as any farmer in the world.

Each of these first-class farmers can articulate — and demonstrate — their philosophies. Both of them wow me with their skill and the ultimate test: the taste of their produce. Coleman's baby carrots trade higher than candies among school kids' lunches in his area. If you have a Coleman carrot, that beats dessert. Canard's veggies burst with taste, which is why Alice Waters uses them. How can both of these innovators achieve similar results with such opposing philosophies? I don't know, but I guarantee each has a lot to teach. I'd be up for an internship with either mentor.

5. *Passion.* Does the farm exude a sense of urgency to a cause bigger than itself, bigger than its players? Plenty of people have accused our family of being workaholics. I don't know if we are or not, but I guarantee you that if we hadn't worked harder than most people, today we wouldn't have a farm attractive to aspiring young people.

"I don't want to work that hard," isn't a statement of endearment to me. I'm well aware of quality-of-life goals, but those goals shouldn't excuse lackadaisical and lackluster approaches to work. American culture now seems to applaud victimhood more than entrepreneurship. We penalize high achievers with extra taxes and broadsides about greed and self-indulgence.

While I appreciate that some do go over the edge, that's no reason to demonize a passionate work ethic. All of us who live and work passionately on causes much bigger than our own are not members of the empire-builders, the dark side, the multi-millionaire CEOs. Passion for money is quite different than passion for sacred causes, like healing the soil and producing such high quality food that hospitals aren't needed. How about that for a goal?

I don't think anyone achieves true competency without passion. That's what drives craftsmen to improve, to never quit tweaking and to keep getting better. A person simply content to do the same thing over and over will never become artisanal.

6. *Efficient.* Does the farm have an always-behind look? Work timeliness is as important in farming as it is in any other vocation. Getting things done is only half the battle; getting them done on time is the other half. Every day has a sequence. Every season has a sequence.

Retracing steps because the logistical job sequence wasn't thought out smacks of incompetence. Spinning around in circles faster isn't the answer. We've all heard about working smarter rather than harder. Planning is the best time you can spend. The Ranching for Profit schools operated by Dave Pratt distinguish between Working In The Business (WITB) and Working On The Business (WOTB). Most of us are so wrapped up in WITB that we fail to practice WOTB, which is the more important key to success.

One of the characteristics of all masters is efficiency of movement. Ever watch a highly skilled heavy equipment operator? They're doing something with that machine every second. They don't spend any time idle, but leverage both forward and backward motion to get the work done.

That's the way expert farmers are. They map out the day and outline the sequence in their heads to minimize lost motion. "Go loaded and come loaded," we admonish on our farm. "Never carry one bucket; always carry two. If you only need one, take two and you'll have an extra one next time. " These are rules to live by, the make-it-or-break-it standards of efficiency. A good farmer has an ease of motion about his work, but underlying it is a sense of urgency. Farming isn't just a walk in the park. Anyone who doesn't move with a sense of urgency isn't worth following.

7. *Financial integrity.* Does the farm pay its bills? Is it held in high esteem in the local business community for on-time payment? Financial integrity precedes all other integrity. While profitability may be related to this skill, it's only a part.

Financial record keeping is essential for being able to explain the business aspects of the economic enterprise. Does the farm just keep all its invoices in a box until the end of the year and then dump them on an accountant's desk?

Does the farm have a good business reputation in the community? I don't think it's unfair for a prospective intern to ask around about the farm's business dealings. Just as people ask for references with resumes, a farm should be able to supply references for their business relationships. Asking for this isn't being nosey; it's just being wise.

8. *Longevity.* How long has the farm been in business? Does it have staying power? I warn both mentors and interns about starting mentoring programs too early in the life of the farm business. A depth and breadth of business experience is key to a good outcome.

No business can do too many creative things at once. Creativity drains money and emotion due to the learning curve. A start-up operation faces newness on almost every level; adding an internship program does not and will not help the situation. Even for old established farm businesses, adding an intern program can be highly disruptive. At Polyface, the farm was in business some thirty years prior to the first intern.

We started with a couple of trial balloons for two years, using experimental, short-term, seasonal interns before launching formally with two official interns. I cringe when I see a brand new farm, only a year or two into the business, jump into an internship program. The only exception to this rule is a new farm operated by a person who has already been farming elsewhere for a few years. The experience accumulates in the person, not the actual physical location of the farm.

Experienced farmers know that each climate and location is different. That adaptation is all part of customizing knowledge to local conditions. Being part of that exciting time can be extra special for interns. But being part of a fledgling enterprise, where even the farmers are novices in financing, marketing, production, and processing is a nightmare. The farmers don't know their own weak links and the interns become frustrated at being jerked around from one unfocused project to another.

Don't even think about starting a mentoring program if you're not extremely comfortable in your own farm. Of course, mentors learn every day too. But there must be a pool of stability that floats the day-to-day operations. Generally, the longer the farm has been in business, the better the intern experience will be for both the farm and the intern. Any farm that believes intern labor is the key to survival is living in la-la land. A farm needs to show proficiency without interns for awhile to establish credibility as a farm capable of working with interns in the fullest sense.

This list is probably not exhaustive, but it gives some things to think about. As a prospective mentor, do you have the respect of others in your field? They don't have to agree with you, but do they at least respect you? As a mentor, you need to answer the question: "If I were aspiring to do what I do, would I want to become me? Am I the person from whom I'd want to learn?" If the answer to that question is yes, you're ready.

On the flip side, you're an aspiring intern. Do your research. You're as responsible for choosing your mentor as the mentor is responsible for choosing you. How is the prospective mentor viewed by peers? Remember, all truly innovative people will have their detractors. But look at the body of weight and make the assessment.

Far more farmers are capable of being mentors than most people realize. I've never visited a farm that couldn't teach me something, even if it was simply about a better gate handle. This list of skill criteria is not meant to be exclusive, but to be validating. Farmers tend to be self-deprecating, a kind of "aw shucks" attitude because society tends to demean their effort and worth. My goal with this discussion is to validate, not limit.

As we close out this chapter, let me emphasize that when I talk about mentoring skills, they're not limited to what our culture broadly labels sustainable or organic farms. Many of the day-to-day things we do on our farm are exactly the same tasks as those done on any commercial farm anywhere. Driving machinery. Backing up machinery. Building fences. Mowing hay. Baling hay. Feeding hay. Planting gardens. Pruning apple trees. Sorting cattle. Sorting pigs. Processing chickens. Loading firewood. Cutting trees. Loading and unloading manure spreaders.

The list is virtually endless, but all these skills have nothing to do with whether a farm uses chemicals or not. These are simply farm skills that reside in the personhood of farmers, one and all. In this respect, the farm class, or farm community, contains a body of knowledge that is transferable anywhere.

To farmers who don't share my passion for compost and ponds, I would simply say that much of what we do here at Polyface involves the same skills you possess. Don't think I'm suggesting you have to have a farm like Polyface in order to have something to offer. We have tractors and other machinery, sheds, plants, and animals that all need control and care just like any other farm. While we may have differing techniques and designs on some if not even many things, we still share a basic universal set of requirements and skills. I don't think anyone has figured out how to back up a trailer turning the front wheels of the tractor right to make the trailer go right.

Whether the trailer contains chemical fertilizer or compost, you have to turn the tractor wheels right in order for the rear of the trailer to go left. That's as axiomatic as gravity and has nothing to do with farmer philosophies.

Now, to you young people looking for internships, looking for a place to start, I turn this advice toward you. Don't be unnecessarily picky. Not only can you learn a substantial amount from any farmer, but you can also possibly endear yourself to him in a way that creates an opportunity you may not envision right now. If you start working with a farmer with whom you disagree on basic philosophy, don't worry about it. Learn all you can. Hammering a nail or building a corral gate is exactly the same on my farm as any other farm in the world.

Take the wheat and spit out the chaff. Just because there's some chaff — and every single internship program will have some chaff — doesn't mean you won't find some delectable wheat. Stay with it. Engage. For sure, internship opportunities with good sustainable farmers are currently hard to get. The pool of good farmers in this category is relatively small and interest is huge. But don't let that stop you. Partner up with almost any middle-aged to elderly farmer and you'll be rewarded with great basic skills. You can take those anywhere and adapt them to any farming philosophy you eventually choose.

In the list of skills I've highlighted in this chapter, notice I haven't picked qualities that differentiate farms by philosophy. I've picked the big picture items. We have some farmers in our community who couldn't be more different than me in philosophy. I've written about some of them in my book *The Sheer Ecstasy of Being a Lunatic Farmer*. But I guarantee you they have tricks, acquired from a lifetime of farming, that would help me be a better farmer. It might be a way to back up a trailer. It might be a way to hold nails in one hand while hammering a nail with the other hand.

I hope that any farmer reading this will realize that we are indeed skilled. We do know how to do a lot of things and we know why we do them. We do have fire in the belly about life, farming, family, and more. And we pay our bills. My challenge, then, to all of us who now realize we have something to offer is to open our hearts to these intern seeds, to plant them and watch them germinate. It's exciting. If we don't, our farm businesses will follow a degenerating trajectory that follows our own physical decline. And we'll grow old on our farms, alone.

I like the advertisements for foster parents that start with the adage that, "You don't have a to be a perfect parent . . . " In a lot of ways, that's the way I see the mentor/intern partnership. The mentor need not be perfectly skilled. The intern need not expect perfect skills. The mentor needs to understand that a lifetime of experience carries an inherent skill set. Imparting those skills earlier rather than later increases the likelihood of a continuing farm. Failure to impart this knowledge and these skills ensures the demise of the farm as surely as night follows day.

The intern needs to understand that an imperfect mentor is better than none. Learning sooner rather than later increases the chance of success before your youthfully enthusiastic, conquer-the-world energy wanes. Earlier learning leverages better than later learning. In that way, education is a bit like money. It carries a time factor, partly because the sooner you begin metabolizing the knowledge, the more it has time to bear fruit.

Skilled mentors are everywhere. Again, I've seldom met a farmer, of any stripe, who couldn't teach me something. Don't get bogged down in finding the perfect fit. Good enough is perfect. Just start. With mutual respect, even some fairly odd couples can find compatibility and enjoyment. Now go for it, both of you: mentor and intern.

Chapter 6

Investing in People

S howing isn't enough. Teaching isn't enough. Working together isn't enough. Pay isn't enough. Mentorship requires something more. It requires an emotional investment. It means getting down and dirty with relationships.

If you're a farm family and you're dysfunctional, don't even think about starting an internship program. I'm going to belabor this point because I've heard too many farmers talk glibly about starting an internship program to get more labor. If you start taking interns to get free labor, you'll be sorely disappointed. What's worse, you'll take down some wonderful young people in your sinking ship.

In short, interns invade your space. They don't ease relationship issues; they complicate them. If the farm couple is struggling in the marriage, or suffering severe disagreements about farm direction, interns will worsen everything. Fast.

Farm families are under perhaps as severe a strain as they've ever been, certainly in America. The Dust Bowl era was bad, but it was regionalized. The pioneer days were rough, but everyone was in the same boat and neighborliness was strong. Today, farm families suffer new stresses. To begin with, they're stereotypically marginalized as unimportant in society. The average American assumes that food magically appears on the supermarket shelves, a tribute to the technologically advanced alchemy of processed, extruded, amalgamated, synthesized, reconstituted, adulterated, prostituted compounds of lifeless unpronounceable pseudo-food.

Not only do most Americans not have a clue where their food comes from, they don't even care. They're far more interested in the latest belly button piercing in Hollywood celebrity culture than the viability of farms. We've entered a day of historically unprecedented marginalized agrarianism. Rather than being the repositories of wisdom, innovation, economic stability and political articulation, they are the butt of hillbilly jokes and red neck derision.

The current local food tsunami and rise of farmers' markets is a welcome trend and shows the cultural pendulum swinging to a more corrective place. "Know your farmer know your food" is a wonderful campaign, as well as the "Buy Local" slogan. With this new-found appreciation and awareness, salt-of-the-earth farmers now frame advertisements and down-to-earth marketing campaigns. I feel sometimes like an ugly stepchild who has now entered a Cinderella world. Indeed, never able to get on a cultural wave, now I feel propelled by Monstro. It's truly exciting. But this is a recent and enjoyable turn of events, focusing primarily on artisanal farms like Polyface.

Unappreciated in society, most farmers have been relegated to the edges of socio-political importance and discourse. Numbering too few to even merit noting in the census, farmers generally have slipped into a defeated, anachronistic mindset. Growing weary on the acreage they love, many encourage their children to seek a better life. Abandoned by an unappreciative society, most farmers have emotionally given up. They plod along because that's what they've always done and they don't know anything better to do. Too old to learn a new trade, they just keep planting, feeding, and showing up at the Ruritan Club until they can't get up in the morning.

It's sad. Really sad. I see it in the faces at the livestock auction barn. I see it in the faces at traditional farm conventions, and too often even at sustainable farm conventions. Oh, they try to put on a good front. They tell stories and reminisce about the old days. Most of the stories happened because lots of people were around ... back then. Whenever a young farmer, boy or girl, comes by, their eyes twinkle as they think of what could have been on their farm. But most of the kids are gone. They went to Dilbert cubicles to work for Fortune 500 companies, put their kids in soccer leagues, and joined the Sierra Club in penance for all the chemicals and plowing Dad did back on the homeplace.

All you have to do is drive out through the midwest, the heartland of America, to see that heart stripped bare and bleeding. Not one in three houses is inhabited. Many sit abandoned and lonely, crumbling, amidst gargantuan fields of genetically modified corn and soybeans grown for animal factories. In a few years, the old houses will go ahead and give up too, like their former owners, and crumble into the soil. Then they won't impede the plow anymore. They will decompose back into the earth, a little spot of fertility harking back to earlier days when the fertile earth sprouted small, diversified farms, communities, tax bases, and livestock shows.

How do we get the life back? How do we create hope in these seemingly hopeless situations? Only families that have vision, singleness of heart, magnanimous spirit, and optimistic hope can realistically establish a functional internship program. It's too hard otherwise.

If you need to work on your marriage, interns will give you a good excuse to put off that hard work. They'll distract you and simply complicate all the issues. You can try to run away to the important task of teaching interns, but in the end, you'll end up with a broken marriage and disrespectful interns. Interns can sniff out dysfunction like a blood hound. Further, interns hate to be placed in the middle of feuds — marital, familial, or otherwise.

To be fair, all of us will have moments of argument within the family business. But it can't be ongoing. It must be talked through, laid to rest, and resolved. Otherwise, it'll show up at the intern's workplace. The interns see and hear things. And then they talk.

Few rumor mills are as functional as interns on a farm. Every problem, every inconsistency, every disagreement, they'll see and begin wondering. Is this family really what it's cracked up to be? Are they for real? Are they using me? Where do I fit in? What's the future for this place? In their quarters, interns can and will fabricate tons of perceptions and misperceptions. Some you'll hear about and some you won't.

If you haven't learned to be transparent and honest in your

own marriage and in the farm family and staff, you'll be called out by the interns. It won't be pretty. Interns will expose your weaknesses. They'll pick apart foolish work, foolish spending, and talk that doesn't match your walk. Too many fledgling farms try to use bravado in their fliers, overstating, over-promising. Interns will grab onto that like a pit bull and you'll see their loyalty slip. You want loyal pit bulls.

When Daniel does workshops about farm internships, he begins by asking, "Are you sure you're ready?" He goes on to explain that every facet of your life will be questioned and scrutinized. Interns will ask you why you go or don't go to church, why you buy this and not that, why you have or don't have a microwave, how you vote, why you bought that kind of tractor instead of the other kind. You'll live under a microscope whether you like it or not. You need to be happy in your own skin, quick to laugh at your own inconsistencies, and ready to admit you don't know it all.

A little humor and a little humility go a long way to create respect. Interns enjoy hearing convictional explanations, to be sure. But they'll examine those from every angle known to man.

A Different Kind of My Space

A word here about privacy. Everyone who starts a farm internship program wrestles with privacy boundaries. The balance between cordiality and formality is hard to achieve. We want the interns to feel like family, to feel enveloped in support and affection. But we don't want them wandering through our home, coming into our private quarters without knocking. They need a retreat space and so does the farm family.

This balance will be different for every family. If the farm mentor family has small children, the need for privacy and space will usually be more than if the family is older. Older couples tend to mellow, becoming more tolerant of invasions. I advise categorically against having the interns stay in the farm mentor family's house. That may sound interesting and romantic, but it won't last long.

The interns need their space, and so does the farm family. At Polyface, we knock on the intern housing door and wait to be invited in. I know we own the building, but it's a recognition, a respect, toward the interns that they have their own space. By the same token, we expect them to knock at our house, and to at least wait on the threshold to be invited in, rather than waltzing in like they own the place. This degree of formality creates a line of mutual respect.

An Investment Beyond Dollars and Cents

Intern programs don't make struggling farms successful. They accentuate weaknesses. They can hurt farms if the personal relationships are weak. They can also break farms if the financials are weak. If you don't know how to make money on the farm, don't assume interns are going to teach you. This is fundamental enough that I'm embarrassed to use ink on the issue. But as I travel and talk to farmers, I never cease to be amazed at the assumption that struggling farms can be salvaged by interns.

It should be a given that farms attractive to interns be stable and successful both emotionally and financially. If you aren't paying yourself a decent wage — and most farmers aren't — interns won't make it happen. Remember, education is expensive. If you can't afford to lose $3,000 per season per intern, don't start. This is why you want to start with just one or two; don't start with several. Work your way in. For sure, good ones will more than make up for their losses. But not every one is going to be a good one. Really stellar ones are few and far between because they don't share your work ethic...yet. That's why they're interns. They don't understand the gravity of the situation. They don't understand the value of tools. They don't have a clue about the value of wasted chicken feed adorning the ground. Or the value of a misplaced hammer or hitch pin. You've got to bring them along.

Interns will intensify the financial strain. This is important because vetting interns is a skill too. When you embark on an internship program, you've just started going to school too. As a novice picker of interns, you're going to make mistakes. I know, I know, in the honeymoon of the euphoric decision when you and your beloved decided that all would be well if you just had some of that "free help," it's hard to imagine a blown tractor engine and upturned pickup truck. You've never done it, so it's hard to imagine anyone else could do it.

But I guarantee you that within a couple of months of starting an internship program, you'll look at each other, in the privacy of your home one evening after a very, very long day, and shake your heads, wondering how it's possible for anyone to be so creative at lousing up such simple things.

Once, I took a roll of electric fence wire and stakes out with an intern. I thought it was simple when I explained that, "I'll put the stakes in and you unroll the wire behind me."

Just so you know what I'm talking about, these rebar stakes have big plastic insulators on them. They're simply a four-foot piece of rebar that I push in the ground with the big, obvious, screamingly apparent bright yellow insulators screwed on at about the three foot level. So I'm heading out through the field counting paces and pushing the stakes in the ground. About halfway across the field, I look back to see how the intern is progressing, and she's coming along, smiling, happy, about 20 yards behind me. The electric fence wire, I notice, seems to be strangely floppy behind her.

Upon closer inspection, I notice that it's not in any of the insulators. I stop — trying to remain calm — and ask her why she's not putting it in the insulators. "You just told me to unroll the wire," she says, as matter-of-factly and innocent as a cooing babe in diapers. Honestly, I can't make this stuff up. Truth really is stranger than fiction.

If you think interns are going to get you out of the financial and emotional doldrums, think again. If you're expecting salvation at the hand of interns, you're not facing reality. You'll wish you'd never heard the word intern. You'll wish the whole notion of internships had never been invented. I know, because I've talked to farmers who begin quivering and shuddering uncontrollably at the mention of the word intern.

In recent years, I'm sure our family has angered many people by turning down their requests for short-term learning experiences. The request usually goes something like, "We're a young family, read all your books, and are ready for farming, but need a little hands on. The four month internship is too long for us, and we'd like to come and just work with you for a week to get valuable hands on experience. We believe this would be just the catalyst we need to embark on our own farming venture. We have a great attitude and know how to work hard, so we're sure we can earn our way. You can't imagine what it would mean to us to come and work and learn alongside you for a week. Please let us know when we could make this happen."

The variations on that theme are as numerous as phrases in the English language, but they pretty much follow that pattern. Many years ago we agreed to a couple of these and found them totally unsatisfactory. In short, nobody — and I mean nobody — can work hard enough to earn their presence. We call these tag-alongs. While we're trying to get our work done, which is actually fairly cerebral most of the time, they're tagging along the whole time pestering us with questions. Think about your workplace and having a precocious family surrounding you all day with questions. Yeah, that's what I thought. Kind of distracting, huh?

The problem is you can't just say, "Hook up the hose over there." They have to know where to hook it up, and then they invariably won't screw it on tight enough, so when the valve is turned on, water sprays everywhere. They will have already walked back to you, so now you have to go and fix it. Or what's more common, one of the other family members is asking a question and you don't notice the leak. You go out for chores in the evening and have no water in the system, or you notice the cows have made a monstrous mud hole in the field, in a swamp that didn't used to be there . . . until some yahoo who knew they could earn their keep hooked up the hose like some kindergartener. The real kicker is that when you finally go over to fix it, you find that the tag-along also dropped the rubber washer out of the fitting so, in fact, you can't fix it until you go back to the shop and find another rubber washer . . . .

Even after we decided to say no to every one of these requests, people still kept asking for the privilege. After a couple of years, we decided to try a new tactic. We'd say yes, but charge $1,000 per person per week. We thought maybe that would at least compensate for the distraction losses. Surprise, surprise. It didn't! We had some eight people who took us up on this new offer the first season, and it was worse than before. The reason was that now we had people who paid good money for the experience, so felt more entitled to ask more questions and work way less. It was impossible.

We discontinued that after one season and have never again

allowed what we call tag-alongs. The pleas are sincere and our hearts break for people. "But we really want to farm and we just can't do the whole intern program. We desperately need this hands-on experience. We're sure we can earn our way." The answer is, "No, you can't. Nobody can."

This is why at Polyface we've gradually refined our learning interface to what fits. We're not a public outfit with taxpayer dollars to pick up the pieces. We're a for-profit business — profit is not all we're interested in, surely — but we can't handle the level of distraction tag-alongs bring. By the way, this is why we've not become involved with WWOOFers (Willing Workers On Organic Farms, an international farm work-learning program). They just don't stay long enough to earn their cost back.

What does fit? First, we have an open door policy. Anyone may come anytime to see anything, anywhere, unannounced. How many farms do that? This openness includes the ability to take pictures, measurements, whatever. As the flow of visitors increases, we've had to learn to just smile, say hi, and keep on walking. I used to stop for every visitor and chat for awhile, but now it's too many. Since we never want to act stand-offish and inhospitable, the open door policy stands. We hope we never have to change it and we definitely don't want anyone reading this to feel like we consider a visit an intrusion. We don't; we don't because we've learned to not feel compelled to spend time with every visitor.

Even among sustainable farms, when I mention this open door policy people gasp, exclaiming, "How can you just have people wandering over your farm?" First of all, we really do want others to take the principles and duplicate them. They work anywhere in the world, with local customization, and will help heal the land, the economy and the community anywhere they are implemented. This is our mission, and we feel strongly about being open to seekers.

Second, we find that if you genuinely expect good behavior and trust people to respect you, they will. To my knowledge, nobody has ever stolen things or destroyed anything. Of course, our farm is covered in people, so we may look as dense as rocks, but we're really spying from behind every corner fence post. We have magnification video security cameras camouflaged in the trees . . . Ha! No, we actually appreciate when families can enjoy our openness, take advantage of it, and realize the flip side of that openness is to grant us a little space in our house.

The other morning Teresa and I were eating breakfast and heard some activity by the kitchen window. I looked over and there was a family of four, faces pressed to the window, hands enveloping their faces to shut out light so they could get a better look, and they exclaimed to each other, "There's where they live, there's where they live." I waved and smiled. Teresa did not.

When you embark on this teaching path, you need to be ready for requests from people who want special treatment, special dispensations. All you have to do is decide what fits, and be willing to firmly enforce the limits.

In the summers, for two years out of three, we offer the Polyface Intensive Discovery Seminars (PIDS), a two-day, six-meal, on-farm intimate learning experience. It's worth coming just for the food, trust me. Limited to thirty people per session, we usually offer three sessions a season, and though we can't accommodate everyone who wants to come, it's a learning situation that fits for us. We do charge for the experience, but that pays us to essentially quit farming for a couple of days and focus all our attention on those who attend. We split the staff so that some people are involved with the students and others are making the farm go, therefore reducing distraction slippage.

Once every three years we host the Polyface Field Day, a one-day extravaganza complete with barbecued chicken, beef, and pork. We cram everything we can into that one day, including a trade show for vendors who supply us. It's a huge event and gives us a reason to clean up the messes and spruce up the place routinely. That fits.

Finally, we offer the internships and apprenticeships. If one of these learning experiences doesn't work for someone, we're very sorry, but all of these have been carved into our lives as the ones that fit. The reason the four-month internship fits is that it's long enough for the interns to compensate their learning with a work trade off. If the internships were any shorter, they wouldn't fit economically or emotionally with our farm or family.

Our observation is that the interns lose us money the first month and a half, break even the next month and half, then accrue profitability the last month. Make no mistake about it, we define profitability as one of the criteria for activities that fit. Just as a college often quits offering courses when too few students sign up to make the class profitable, we define our opportunities according to what creates a positive bottom line. This isn't mercenary; it's survival.

As a farm family, we pour our lives into these young people. We want the experience to be as memorable and positive for them as it is for us, and we've generally found them to be unbelievably forgiving. We've not always had accommodations completed on time. We've not always had job lists written down on time. We've not always explained things the best. We've not always chosen the right people, sometimes throwing the group into difficult relational dynamics.

Any time you deal with people, relationship issues surface. Everyone comes to the internship with preconceived notions — including the mentors. We each have expectations, and sometimes no matter what each side does to reduce confusion, it still happens. That's life. The key is to be open, to be quick to apologize. A quick apology can cover a multitude of problems.

We require everyone, including ourselves, to take responsibility for actions. When the work environment is as intimate as a farm internship, things can get said or done that will grate, affecting the relationship. Mood swings and irritations happen. As the farm mentor and ultimate leader of the group, you need to be ready and willing to shoulder the emotional equity that will keep things pleasant, trusting, and appreciative.

When an intern complains that, "You didn't explain it to me," that's not the time to snap back with, "How can you be so dense?" Interns come in good faith. I don't know any intern who has ever come with a chip on their shoulder or anything except expectations to learn and enjoy the season. Whether you thought you explained it well enough or not isn't the issue. The issue is that the intern, the student in this case, didn't get the message.

I remember taking an education course once where the theme was: "If the student hasn't learned, has the teacher taught?" It's a powerful question, and mimics in education settings the formula for work: mass times distance. In other words, according to physics, if I push on a wall all day but it doesn't move, I haven't done any work. I could grunt and sweat and become exhausted, but if the wall doesn't move, I haven't worked a lick. Isn't that fascinating?

In educational terms, then, if I give instructions all day but the students don't understand any of them, have I taught? No. Part of creating emotional equity with the interns is to assume that they want to succeed, that their feedback is important. If you want them to trust and respect your instructions, they need to know that when they level with you about being confused, you'll take them at face value too. You can't have it all one way. You can't have a good working or learning environment if the student is routinely considered an imbecile — even if he is — and the teacher is always right. That's not the way to create healthy relationships.

One commonality among interns is that they're far more incompetent than they realize. But as the mentor you have to be aware of that and compensate for it. A good internship program requires the mentor to be willing, able, and happy to pour lots of emotional equity into the arrangement. Otherwise, the interns will get discouraged, disloyal, and eventually vindictive. That's not a good thing.

Fortunately, showering the interns with emotional equity always yields incredible returns. This isn't a net subtraction from your life. On the contrary, it's a glorious net addition. But you, the mentor, as the leader, need to be the pace-setter. The interns will respond in kind, but someone has to start things. That's you, the mentor. Don't sit around expecting them to fill your emotional tank. Instead, make up your mind to pour in a heap of emotional support up front, and it'll come back in buckets.

What does emotional support look like?

Praise. Watch for little things — sometimes that's all you've got — to compliment them. If all they hear is grumbling and "it's not good enough," they'll get discouraged and then angry. Be liberal with praise at every opportunity. You'll need to draw on that goodwill when you need to be stern about things or when you express displeasure. If you've praised routinely, then when you jump them for lolly-gagging, they'll jump to please you. Most people prefer praise to complaints. It's okay to complain, to voice your displeasure, but you earn the right by having heaped on some praise beforehand.

Gratitude. I know these interns are costly, and they're supposed to do the work you've demanded. That's their reasonable service, right? Say thanks anyway. Gratitude begets gratitude. Show your appreciation for a job well done. This is the best pay in the world, and keeps them from feeling like you're simply an extractor. Even though the interns owe the work and effort, they need to know their sweat is appreciated.

Apology. No matter how well you try, you'll say an unfit word. You'll make an incorrect assumption, or give confusing instructions. Sometimes you'll change your mind and realize your original plan was wrong. In these cases, be quick — even eager — to apologize for the mistake or unfit word. It's bound to happen, so just prepare yourself emotionally and look forward to the opportunity to apologize. It'll be so striking to the interns they'll remember it forever.

Magnanimity. I thought about using the word hospitality here, but I wanted to go beyond that. In this context, magnanimity means being willing to take the dirtiest job. This means pushing yourself when you're tired to stay out there with the interns and not retreat to the house and a cool lemonade. It means waiting for a drink of water until everyone can get a drink of water. It means not taking advantage of your privileges and making the interns feel like dirt. This whole relationship is about way more than money and getting the work done, although both of those are important. Ultimately it's about creating an environment that encourages each individual to exceed their potential and carry that enthusiasm into every area of their life. Be an encourager, a motivator. Having a magnanimous spirit where there's always plenty rather than presenting a persona of scarcity (like Stephen Covey's *7 Habits of Highly Effective People*) will fill everyone's emotional tank.

At the end of the internship, these young people will leave carrying your spirit and attitude. Believe it or not, they'll catch the unseen heart of your life. Yes, they will leave with technical understanding and a great work ethic. They'll leave with friendships and experiences. But the most compelling thing you can give them is the passion of your soul, the unseen driving force of your spirit that makes you want to see, and cultivate fields of farmers.

Especially for Interns

Chapter 7

Cost

Perhaps the most misunderstood aspect of farm internships is how costly they are to the farm. As I've already noted, education is expensive. In our case, we don't charge for it; we just ask for some work — okay, a lot of work — in return.

For the sake of this discussion, I'll use interns and apprentices interchangeably. Later I'll discuss the technical differences between the two in the Polyface program, along with the differences in remuneration and responsibility.

While we do pay a stipend, that's not the major cost in running an internship program. The major costs are in broken machinery, lost tools, and work slippage. That includes work that must be redone because it was done incorrectly; work that didn't get done when it should have; and things like gates left open or the electric fence unchecked. Open gates and dead electric fences usually mean animals where you don't want them. Sometimes lots of animals, and sometimes very far away from where you want them. More about that later.

I could list, in accounting formality and precision, the number and types of these instances. In learning's crucible, good intentions don't always yield good results. In order to convey what will happen in an intern program, I've selected a smattering of real-life stories to explain the cost of apprenticeship. Stories are far more entertaining than accounting lists.

The High Price of Lost Mementos

This cost came home to me when we first launched the program and I began to feel like a visitor in the farm tool shop. In the beginning, my Dad was the king of the shop. He had tools that he had acquired during his lifetime. Dad was always inventing things, so he spent a lot of time in the shop. Because he encouraged me to participate, I spent a lot of time in the shop, too. Many of my earliest and fondest childhood memories involve the shop.

We had a little arc welder that Dad affectionately called the "Buzz Box." That served for most of our little fabrication projects. We had a big one out in the equipment shed that Dad rigged up to run off an old two-cylinder Wisconsin engine that he scavenged off the Case baler that came with the farm. When he and Mom bought the farm in 1961 — after fleeing expropriation in Venezuela — it came with five pieces of equipment: a 1947 gray and red Ford 8N tractor (the wooden box it came in was out by the barn and repurposed for a pig house), a Case square baler, a 3-point hitch-mounted cycle-bar hay mower, a hay wagon, and an old horse-drawn rake remodeled with a hitch to pull with a tractor.

That represented the sum and substance of our line of machinery, as farmers are wont to say. At the time, the farm was not a going concern. But Dad was a genius, and very quickly set to work designing a portable electric fencing system, portable shade mobile, a portable veal calf barn, and other technologically advanced infrastructure. To do all that required tools, and these tools became my friends.

As Dad's go-fer I became intimately acquainted with his designations: red-handled pliers, ball-peen hammer, nickel-plated pliers, yellow-handled screwdriver, claw hammer (we only had one), and one-handed sledge. I fetched these like a dutiful assistant hands instruments to a surgeon: scalpel, suture, gauze. I knew each one's place and how it felt in my hand. We never had a television (still don't) and not many toys. Dad and Mom didn't have money to dote on us with gadgetry from town — in that era, children mostly played with blocks, cardboard boxes, and string. We didn't have talking toys and robots. How deprived, huh?

At any rate, these tools became part and parcel of my personhood. They identified a bond between father and son. I'll never be as mechanical as Dad, nor as clever, but as I grew into teenage years, these special, named tools were as beloved as cousins. Being from the Midwest and coming to Virginia via a decade-long farming stint in Venezuela, our arrival in our new community in Augusta County, Virginia, didn't include relatives. One of the things I feel most blessed about today is having four generations on the farm.

But back then, these tools were my friends and extended family. I learned how to use them — or rather, how Dad used them — and faithfully kept them in their rightful places. We had two crowbars. One was round and a little bit longer than the other, which was octagonal. The digging iron, used for prying and loosening dirt in post holes (and then tamping it tight around the post) was a piece of automobile axle. Dad created the sharp wedge on one end by heating the bar and beating it out into a flared wedge with a hammer, like a blacksmith.

As I began taking on more responsibility for the farm, I began adopting these tools as my own. They evoked a strong emotional bond and were an extension of myself. People who don't work with their hands have a hard time imagining how personal tools can be. I suppose a distant equivalent would be the way 20-somethings feel about their iPods. But that still doesn't come close, because all of us know the iPod will be obsolete in a couple of years. It has "limited life" stamped all over it.

Perhaps a better equivalent would be that favorite book that Grandma read. Or perhaps the smell of Grandma's house. Maybe your favorite food that Grandma fixed.

But the life of these tools endures. For the most part, they never wear out. With them, Dad showed me how to splice a wire. He showed me how to shape out the edges of a rivet so it would swell into the hole and then taper over to grip the edges of the plate. Before the days of cordless drills and cordless screwdrivers, we had a brace and bit.

I became one with that brace and bit, building my first pastured poultry shelter with my sister when I was about twelve years old. Before the days of drywall screws and self-tappers, I had to drill body holes and then, with flat-headed wood screws, pull the joint toward me while pushing the brace and bit in my gut, cranking it home before stripping off the slotted head. Oh, the memories. Although this pull-push technique is still required, today we use cordless drills instead of the brace and bit. The brace requires far more overall balance and technique since the operator has to provide the motive power, not an electric motor.

Then when our own children were born, they began accompanying me to the tool shop. We still didn't have television. We still didn't buy many toys. Are you getting the picture here? These tools spanned generations. When Dad passed away, the tools were monuments to his memory. Yes, tombstones are okay, but they don't hold a candle to his tools. Dad's hands shaped around mine, holding, guiding, helping. Now mine, shaped around my children's hands on the same tools, held, guided, and helped them. My, if tools could talk, the stories they'd tell.

My favorite tool was Dad's set of three sharpening stones. As a journeyman pattern maker at a Chrysler automobile plant in Indiana, he made his own wood gouges, chisels, and shaping tools. Then he made a beautiful wooden toolbox to store them. To keep them sharp, he had three stones, two encased in hand-made wooden boxes with his name etched in a silver nameplate on the lid. The first was a coarse stone. The second was a fine-grained India stone.

The third stone was a fine white stone with beveled edges to sharpen the concave and convex gouges. It was encased in a fine metal case, hinged and tapered to fit the wedge-shape of the stone, and topped with his name etched into the metal case. I spent many mesmerized hours as a child watching Dad sharpen knives, axes and the beautiful professional machete he had managed to get stateside when we fled the junta in Venezuela.

I've never seen such an instrument on any farm in America. It had fine steel and would keep an edge as sharp as a razor. Because it was heavy, you could swing it for hours without fatigue — the tool did the work. That's one of the distinguishing differences between cheap and high quality tools: good tools are heavy, and work for you. Who wants lightweight shovels and spades? No, you want something that keeps going when you ram it into the dirt. Something that will carry inertia through the stroke.

When I was about twenty years old, for my birthday, Dad gave me his sharpening stones. It was a rite of passage. Of course, through my teen years, I had practiced and practiced sharpening my farm knife, hunting knife, and then axes and hatchets. I can still remember when he felt the edge, grinning, and said: "That's as good as mine." Not long after that, on my birthday, he transferred ownership.

Having a son without his mechanical or woodworking prowess may have been a bit of a disappointment to Dad, though he never showed it. We had a mutual understanding that my eighty-seven degree angles weren't what he'd do, but he never disparaged my efforts. Bless him. With this background, in hindsight, I think I have a deeper appreciation for why he gave those stones to me. They represented the one success of his mechanical transfer to the next generation: sharp knives. Perhaps he realized that was the best of his mechanical ability that he'd ever transfer to me. Ha!

So imagine the day when I was in the tool shop with an intern who haphazardly picked up that white beveled stone, pulled it out of the case, and then accidentally dropped it on the concrete shop floor. Go ahead, contemplate that for a minute. Dad had been gone a decade and I'd just started helping Daniel to learn blade sharpening on it. Now it was forever broken, two pieces, unusable. Today the stone is still in its place out in the shop, in the engraved case, broken. Did someone say something about the cost of internships? It's money, sure. But a whole lot more, too.

Within a year of starting the internships, every one of those tools was gone. The only thing we still have is the toolbox with Dad's gouges. It's too heavy to move and too sophisticated to understand, so everybody leaves it alone. But if it can be moved, and if it can be used, it's lost. Lost out in the woods. Lost in the fields. Just another interesting piece of iron fashioned into a tool. You can buy boatloads of them down at the hardware store.

Wanna make a bet? I confess that I had some deep emotional trials to work through, to come to terms with this new reality in my life. Interns don't see your tools as links to the past. The tools have no sentimental value, no memories. Their loss doesn't bring tears. They're just hunks of metal to be used, abused, lost and replaced. You can get new ones any day of the week down at Lowe's. Folks, that's a real cost of internships. Mentors need to understand that lost tools are a reality. The more hands you have going through the shop, the more tools you'll miss. None of this is intentional; it just is. Kind of like kids and dirty hands.

We have a wry way of describing tool attrition at Polyface: we're seeding the next iron ore bonanza. In a millennium, when the world is desperate for iron, someone will come along our neck of the woods with a sophisticated metallurgy monitor and discover a mother lode of iron. The accumulation of decades of lost tools will build up the biggest iron ore strike ever discovered, yes siree. Wish I could be around to see it.

I'm not telling this story to scare mentors away. This book is about being forewarned. I've gotten over it. Now, we have four digging irons, six crowbars, ten one-handed sledges, eight claw hammers, and diamond sharpening plates. Instead of one mattock, we have ten. The machete? Gone. Now we have a dozen, all cheaper American imitations of the true blue tropical multi-use tool. Now, I view buying pliers by the packet and tape measurers by the box as a small — well, maybe not so small — price to pay for the joy of being surrounded by exuberant youthful zest.

You see, when you put it in those terms, it doesn't sound so bad, does it? Would you rather grow old like a grasping curmudgeon, by yourself, all alone, tottering around your old tools like an outdated, archaic has-been? Of course not. So get over it — it's just a cost of internship. And no matter what you do to try to protect yourself from lost tools, you'll never hold onto them all. So release it. Let it go. The alternative is too devastating to contemplate.

The High Cost of Reinventing the Wheel

Now to broken machinery. Did I say education was expensive? We had an ugly farm truck — I kid you not, we actually entered it in the county fair's "Ugly Truck Contest." We would've won hands down had it not conked out, right in front of 2,000 people, in the middle of the 100-yard track. The rules required that the truck run. I drove it over to the fair and when my moment of fame came, it roared to life and I drove it up onto the track. In ten yards the engine died. I set my jaw, coaxed, and used every mechanical trick in my arsenal, but couldn't get it started again. Meanwhile the announcer in the grandstand is asking the crowd, "Can he get it going? Oh no, looks like this one won't make it. Time's running out." The crowd is cheering, "Come on! " No question, I had the ugliest truck by a landslide. Goodness, some of the other entries even had license plates. But this truck, this beast, this incorrigible, ornery, hunk-of-junk decided to have the last laugh.

"I'll teach you to put me in an ugly truck contest. Your faithful servant. How many tree stumps have you run me over? How many mud holes have you slogged me through? And you're going to make fun of me in front of these people? No way. You're not getting off that easy, you ungrateful door-slammer."

Disqualified, I was mortified. The tow vehicle came and pulled me off the track. As soon as we got off the track, I turned the key one more time and the engine roared to life. That cantankerous truck had it out for me. You see, we had already arranged its sale prior to the entry into the fair, so this was its final voyage under our care. I think the truck had one final "I'll get you" up its sleeve — a strategic tantrum in the middle of the competition.

A couple of years before this final journey, we needed the truck one morning and it wouldn't start. Some day I intend to write a book about our rag-tag assemblage of buildings and machines because often the farmer's ingenuity at squeezing one last breath out of machinery is the difference between profit and loss. At Polyface, we're the last stop for machines. The only place to go after we're finished with them is the scrap metal yard.

Nobody wants to come and buy used equipment from Polyface. That's the way it was with this truck, the ugliest truck in Virginia. A composite of assorted pieced-together parts, it was the most functional truck design we ever had . . . when it ran. But the problem was it was fickle. As illustrated by our contest experience at the fair, it had a mind of its own. You never knew if it was going to start or not. Of course, the fact that the battery was strapped in by heavy bungee cords because the platform had long ago rusted into oblivion didn't help stabilize the terminals to insure good electrical contact.

The cam-operated fuel pump had long ago been replaced with an electric fuel pump with on-again, off-again electrical connections. Because of all this make-do rigging, the fuel pump ran in fits and snorts rather than a purr. The starter had been rebuilt probably half a dozen times. The ignition switch hung in the shattered dash board like a web of tangled tentacles. Any one of these could, and often did, malfunction. What I'm getting to is that pull-starting this beast was a routine occurrence.

But once it was running, the truck was perfect for our abusive requests. We'd overload it until the leaf springs bent down U-shaped and the bed rubbed the tires. It was a four-wheel drive, with a hi-lo range and a dump bed on the back. It was almost a tractor and to my knowledge never got stuck. One particular morning it wouldn't start so I told the intern to hook up a chain to the front end while I got the tractor to pull start it.

Now, when I say "hook up a chain," that carries a tremendous amount of pre-understanding. I realize that now. But back then, I assumed anybody in their 20s would know how to hook up a chain for towing a vehicle.

When you look underneath a vehicle, there are a couple of places where you might want to hook a chain and lots of places you wouldn't. For example, you wouldn't want to hook the chain on the tire because it has to spin. You wouldn't want to hook the chain on the drive shaft because it has to spin. You wouldn't want to hook the chain on the idler arm of the tie rod . . . Oh, but if you're an intern, you just might.

Yes indeed, see that spindly little rod under there with all that room above and below? Perfect place to flip that chain around and hook it up. Yessirreee. Hook up a chain? Easy as pie. To get the full import of this fateful decision, realize that this was an old truck and very unlike the tight front ends of today's trucks. Today the manufacturers imbed towing hooks in the bumper because modern engineering has closed up many of yesteryear's spaces where tow chains could be hooked. You know, all the open areas around the engine mounts that allowed you to easily replace a starter or alternator. Lots of space, like around the ends of that big leaf spring, dimwit.

Anyway, the intern hooked up the chain to the truck and I backed up the tractor to receive the other end of the chain. I put him on the tractor since it didn't take much skill to just drive the tractor forward. I climbed in the truck. The skill was in manipulating the clutch, gear shift, accelerator, choke cable, and listening. Ah, now that's a novel thought. Listening to see if the engine turns over, coughs, purrs, or whatever it's going to do.

I instructed him what gear to put the tractor in and off we went. After a few yards I popped the clutch and the engine immediately roared to life. But something funny happened. At the exact moment the engine roared to life, I could no longer steer the truck. I hit the brake, got out, and walked around to see what had happened. Remember that little skinny piece of idler arm on the tie rod? You guessed it. The intern had wrapped the chain around it and we had just pulled it into a V-shape. We now had a grossly pigeon-toed truck. Both front tires looked like they wanted to turn toward each other. It was the most cross-eyed front end you've ever seen. The right tire turned left and the left tire turned right.

I looked at the intern. The intern looked at me. "Did I do that?" seemed the appropriate response. Instead, he just gazed at the redesigned front end, as if it were some piece of modern art. "Well, that looks different," I said, ruefully. He still had not a clue what he'd just done. We crawled under the truck and I pointed out the engineering of how the idler arm is supposed to work. For starters, it needs to be straight. Yeah, right.

Unable to steer the truck, and of course unable to proceed with the project I had in mind, the only recourse was to ease the truck up to the shop where I could heat the idler arm with a torch and try to bend it into some semblance of straightness. I fired up the acetylene torch and commenced applying heat; shortly the rod glowed orange at the apex of the V. I pushed back but it wouldn't give. I hated to take the time to jack the entire front end off the ground, so elected instead to hook up a come-along and winch the bend out of the rod.

I hooked a chain to the rear truck hitch, ran it underneath the vehicle to the front, and hooked another chain to the bend, splicing the two chains with the come-along, also known as a hand winch. I reheated and pulled on the winch lever. It moved nicely, and within a few strokes I had that rod straight enough to be functional. We never did get it perfect, and never did replace it, but on the farm the seriously unaligned front end didn't create shimmy problems since we never went faster than about 15 miles an hour. We put our tools and torch away and headed out to the woods, only an hour behind schedule. Just one more lesson learned. And now, back to work.

Road Grader vs. Pickup Truck on One-lane Road

We had a herd of several hundred cows over at one of our rental farms. Every day we took an intern with us to move them, set up cross fences, move the water trough, and check things. After several days, the intern said he felt completely comfortable to go do it himself. Great, good opportunity to delegate and test proficiency.

He drove over the next day, excited that we trusted him. We, of course, shared the excitement and accomplishment. About thirty minutes later, we received a call. "The truck isn't too bad. It went up on its side but a wrecker pulled it off. I just got charged with reckless driving. Could somebody come and pick me up from a shop where they're going over it to see if it's drivable? Oh, I moved the cows and they're fine."

How would you like to get that phone call? Well, good for moving the cows. That's what we were concerned about. We really didn't expect that you'd roll the truck. What happened was that as he approached the farm on the dirt road, the highway department was running a road grader. The newly graded soft gravels rolled as he swerved to avoid the grader and he went up on the road bank. The truck rolled over — actually, pretty gently — on its side and sat there . . . on its side. He climbed out the passenger side window, which was above him, looked over the truck, and decided it was drivable.

He had a chain in the truck and begged the grader operator to pull him back up on his wheels, but of course, the government protocols required notification of the state police. So the intern had to wait until the police arrived, even though he knew the cows needed to be moved. He actually tried to call us the moment it happened, but we were all out doing things, unable to hear the phone. After all, we had no need to worry. The intern was taking care of everything.

The officer cited him for reckless driving and called a wrecker. The intern ran and moved the cows, then came back to ride in the wrecker to a shop. The shop declared the truck drivable, so Teresa and I went to pick up the truck and the intern. I drove the truck home and they followed in the car in case I had problems with the truck. The intern showed tremendous maturity and paid for the whole thing, even though we did not and would not have asked for compensation.

But looming in the distance was the traffic court date. I went with him to vouch for his character and plead for reduced charges. The judge bought it and reduced the charge to failure to maintain control, which was only a fine of about $50 and wouldn't show up on the intern's driving record. Yes, I happily took the morning to put on a suit and go to court with him. That's what you do for your kids, right? Kids are wonderful, but they aren't always cheap.

What Pole?

Here's one to put a smile on your face. Daniel was in the tool shop, which is part of our extensive equipment shed complex. He dispatched an intern, who was plenty trained and competent, to hook up the pig feed buggy to go feed pigs. The next thing Daniel heard was splintering wood. He turned his head just in time to see the roof of the equipment shed sag down, newly severed from its supporting pole and braces.

The intern had neglected to watch behind him as he pulled out of the shed and caught the corner of the feed buggy on the shed pole. Even though he wasn't going very fast, the fully loaded feed buggy was heavy enough that the inertia carried it through to finish splintering the pole. Daniel and the intern spent the rest of the day propping up the shed roof, pulling the old pole stub out of the ground, prying off all the braces from the top of the pole, digging out the hole and installing a new pole, then reconnecting all the braces and girts.

In an extremely measured and controlled voice, when it was all over and they were taking tools back to the shop, Daniel said to the intern, "Don't you ever, ever, ever, say anything about not being paid enough." Yep, free help alright. Just the other day, at the dinner table, one of our interns expressed the recently-acquired epiphany that, "You have to notice a lot of things when you're driving the tractor." Indeed.

Was There a Car Back There?

A group had come out for a farm tour and, as some visitors are prone to do, parked their cars somewhat haphazardly behind the house. One car in the group, a brand-new Lexus, stood out. One of our interns was doing something in the vicinity with the tractor, and inadvertently backed into the brand new Lexus. It wasn't a big dent, but in a brand new Lexus, it looked inordinately large. This Lexus didn't have 150 miles on the odometer. I mean, this baby was brand-spanking car lot new.

The owner, being a very understanding fellow, assured us that were it any car except a brand new Lexus, he'd forgive, forget, and all would be well. But this was a brand-new Lexus. Only owned it a week, you see. You know, we really needed to fix it.

The little dent cost us $875. We paid for it. Cost of doing business. Cost of interns. Wouldn't you help out your son if he needed it?

Something Old . . .

Back in the early days, we ran extremely used equipment. But as the farm became more profitable — long before interns — we began investing in better equipment. I remember the first new tractor we bought. Dad would probably roll over in his grave if he knew I ever bought a new tractor. The old Case baler that came with the farm eventually went to the metal recycling graveyard — actually, the engine went onto a welder and the hitch and rear chassis went under a discarded dump truck box. We still use that trailer. The axle and hitch are what, maybe seventy years old? This trailer being much larger than a hammer or screwdriver, hopefully, it's too big to lose.

We bought an old hay loader, which is essentially an inclined plane that picks up a windrow of hay and shoves it up a sheet metal incline, dropping it out about ten feet in the air, above the bed of a hay wagon. Dad invented an amazing dump trailer twenty feet long with winch and cable that was closer to pulling itself up by its own bootstraps than anything I've ever seen. The idea was to dump the hay in the barn and have it stand up like a bread loaf on end. Except the hay didn't cooperate.

As the trailer approached vertical, the loose hay just sloughed off in a squatty ten foot pile instead of a nice compact twenty-foot bread loaf. Dad admitted defeat on that invention and we invested in an ancient baler — an old New Holland with a wooden wad board. We limped along with that for a couple of years. It was our baler when Dad passed away in 1988.

Parts were hard to find, so when the local "For Sale" paper had a mate for cheap, we bought it for parts. We scavenged parts off it for a year or two to keep our other one going. One year, on July 4th, we had taken the children in to the Happy Birthday USA Parade in Staunton and returned home to bale two windrows around one field. Although we were going back into town for a family picnic (did I mention that Teresa is related to half of Augusta County? — compensatory relatives for my growing up without any cousins around), we had plenty of time to get these two windrows up and the hay wagons pushed into the barn in case it rained.

The plan was to stay in town after the family picnic until fireworks that evening. We had three hours between coming home from the parade and having to go back in for the picnic. Plenty of time to bale two windrows onto a wagon. We came home from the parade at noon, changed clothes and jumped onto the tractor which I had left parked in the hay field at the first windrow. We started out, Teresa driving the tractor and me stacking hay on the wagon. But within about a hundred yards, the baler malfunctioned. I got underneath and found that we were missing a bolt. I ran back to the shop and got one I thought would fit. Precious minutes were ticking away. Instead of taking the time to run back out to the field on foot, I decided to jump in the pickup and drive out. I drove out there like a crazed banshee and put in the bolt. Good. Up and running.

We went another hundred yards, and things suddenly stopped working again. Another broken part. This time I had to run back to the shop, find a little piece of flat bar, cut it with a hacksaw, bore two holes in it to fabricate the part. Back out to the field in a cloud of dust, ripping and roaring. Scaring the kids. Goodness me.

Amazingly, the fabricated piece worked. Okay, back up and running. We made it two-hundred yards that time before another bolt fell out. Run back to the truck, fly back to the shop. Almost took out that gate post peeling around the corner. Scrounge up another bolt. Back to the field. Screech to a stop — grass doesn't provide very good traction, so you can slide all four tires twenty feet without too much effort. This is all part of teaching the kids, waiting patiently on tractor and wagon with mom, how to drive when they get older.

The bolt worked. Start the tractor. Engage PTO (Power Take-Off). Up and running . . . again. We went about another two-hundred more yards — by this time we were within striking distance of the end of the first windrow. If all held together, we could finish in two hours what we'd planned to do in about twenty minutes. Suddenly, with a mighty crash, then a thump, then a tinkle-tinkle, the whole wad board and the guts of the baler disemboweled to the ground. It was as if the baler vomited all its mid-section parts out on the ground.

Completely exasperated, I said (probably yelled) to Teresa, "Shut the tractor off. We're parking it right here. I quit." By this time the kids were discomfited and I skulked back to the pickup where I'd parked it from my last screeching halt from dealing with the third breakdown in six-hundred yards. I drove to the tractor, the rest of the family piled in the truck — gingerly, I might add, not knowing what kind of driver I'd be — and we tore off, just shy of Richard Petty style, back to the house to change clothes and head to the picnic.

The next part of the story is captivating if you know my dear, frugal, watch-the-pennies, beloved wife, joy of my heart and power of Polyface success. She announced, "Let's stop by the equipment dealer and look at balers." You can't imagine what such an utterance meant to me. That was like a "sic 'em" to a German Shepherd guard dog. She who is the secret of all our financial success; she who never shops, but hunts instead; she who mends work clothes and cans hundreds of quarts of garden produce each summer. Yes, she said to look at a brand-spanking-new baler. I mean one with the paint still on.

Not willing to let this opportunity escape, and completely aware that we didn't have the money for a new baler, but instinctively and completely trusting the financial prowess of my bride, I jumped into clean clothes and we headed to the local New Holland equipment dealer. It was closed.

To roars of laughter, Teresa still recounts this story with the punchline, "It's a good thing it was the Fourth of July, because if that place had been open, I'd have bought that brand-new baler that we couldn't afford on the spot." Whew! Close call, that one. That's my gal, always trying to please. We worked out an arrangement to borrow our neighbor's baler for a couple of years after that and then ended up buying a high volume John Deere that we still use today.

Oops, I Forgot the Tailgate

We know what running dented, junky-looking trucks, tractors, and machinery is like. As the farm has become far more fertile and more productive, we've upgraded. A few years ago, I saw in the same local "For Sale" paper (got to quit getting that thing) an ad for a diesel F-350 four-wheel drive with some age but low miles — "Must See To Appreciate." Teresa and I drove over there and it *was* a beauty.

The owner had bought it to pull a little cattle trailer. He had a small farm and had gotten interested in longhorn cattle. He went to some purebred shows and bred up a couple of bulls. After a few years he'd lost enough money to convince him to sell both the truck and trailer. Shed-kept, waxed and polished, the truck looked like a showroom model. So did they. A handsome middle-aged couple, their home was impeccably furnished and accessorized. Spotless. The truck was the same. Teresa and I test drove it and she agreed it would be a great buy. The price was right and it would become by far and away the nicest truck we'd ever owned. When you've made do — what's the phrase, "use it up, stretch it out, make it do or do without?" — as long as we have, to actually have a nice truck is like finding the perfect wedding dress.

I try to keep the right perspective on these things. They're only things. A wonderful, well-to-do friend used to say to me, "It's just stuff." Yes, but we were sure feeling special in this sparkling, purring, snazzy diesel truck.

We drove it home and it was an instant hit. Daniel and Sheri thought it was pretty enough to take to church. Everyone admired it. I think it might have been the prettiest truck on our road . . . for awhile. Until an intern unhooked the gooseneck trailer and forgot to put down the tailgate. Crunch! The beautiful truck instantly acquired a T-boned tailgate. Did we get upset? No. It's just part of the territory.

You see, there's no free lunch. The intern felt badly enough. No use making a bad situation worse. You can't be surrounded by youthful foolishness and inexperience — and enthusiasm and strong backs — without a few disturbances along the way. When things are bent out of shape, it's just not worth getting bent out of shape emotionally, too. That's the cost of free help.

This chapter will be a whole book if I continue with these types of stories. They are the stuff of legends. An intern actually started compiling some of these memories into a book he teasingly titled, *Legends of Polyface*. Just think how boring life would be without these muck-ups and mishaps. I mean, who wants to live predictably, where everything bad that happens is your own fault? It's a lot more fun when you can share the mistakes with others. And these interns can do far more crazy things than you could ever imagine. So consider it just living in Comedy Central. How many people get to do that?

The Trials and the Errors

Finally, we come to slippage. That's when work is done incorrectly and we have to pick up the slack. Some things can only be learned by going through them. None of us thinks we can louse things up that badly . . . until we actually do it. Daniel and I constantly stress the gravity of the situation. Don't assume all is well — check, re-check, double check.

Probably the best story to illustrate slippage happened when we dispatched an intern to the slaughterhouse with a couple of steers and a cow. Everything loaded fine and he headed up the road for the thirty-mile drive. I was walking across the yard when Teresa called me from the house to come in immediately for a phone call.

I could tell by the sound in her voice that it was serious. I ran into the house and she handed me the phone. It was Tommy May, the owner of the slaughterhouse. Tommy is an institution. At the time he was probably seventy-five years old and he and his wife, Erma, had owned the abattoir for nearly forty years.

Have you ever studied the hands of meat cutters? They are massive. Tommy was a gentle giant. He hadn't gone past eighth grade, but held onto a small community slaughterhouse when countless other operations of his size were going under, collapsing under the weight of prejudicial food safety regulations. He spoke extremely slowly, and exuded a tempered wisdom that only comes from decades of boot-strapping entrepreneurism and the kind of hard work people today consider abusive. In his gentle drawl, his voice came through the phone saying, "Well, Joel, we've had a bad day. Your boy got in a little trouble."

He then proceeded to explain that the intern had failed to back up to the corral loading alley squarely. When you back up a cattle trailer to load or unload, you need to fit the rear end snuggly against the corral opening so that no animal can squeeze between the back of the trailer and the corral and get away. He had parked kitty-cornered but assumed it was okay since he'd done this several times without any problems. The animals always stepped off the trailer, walked down the corral alley, and into the pen. He followed them leisurely, closed the pen gate, and headed for home.

Did you catch that on this load he had steer and a cow? That's significant. Cows are much smarter than young steers. Much more cagey and wily. They have years of experience. And this one was not only a cow, she was half Brahman, which means she was much smarter than most. He let them off like normal and they all headed down the alley. He was probably about fifteen feet behind them when the cow arrived at the pen. As the docile and ignorant steers crowded into the pen with her, she decided she didn't want to be in the pen.

In full flight mode, she wheeled around and came back out into the alley, headed straight for the intern, who, as I said, was nonchalantly following the group about fifteen feet away. She came at him at a full gallop, knocking him aside like a matchstick, and headed for light. The light that was around the side of the trailer. In one leap, she jumped through the crack and headed out into the slaughterhouse parking lot.

The intern picked himself up off the alley floor and took off after her, gum boots sloshing across the parking lot. The slaughterhouse is located in Harrisonburg, a town of 40,000 people. Not out in the country. Right in the middle of a bustling town. The cow headed out onto the street in front of the slaughterhouse, turned left, and headed toward one of the main city highways. The intern couldn't think of anything else to do but keep those gum boots pumping as fast as his little legs could go, trying to keep her in sight.

Almost immediately, two city cops on motorcycles showed up. They gave chase and scared the poor cow to death. She went completely crazy, berserk, even. With cow nostrils snorting and sides heaving ahead of them, the addled cops called for reinforcements. So the SWAT team arrived, big guns and all. They finally got her cornered in a little lot with fencing around it and the two sides started a standoff. She's facing them, almost tuckered out, but certainly not willing to do anything a human wanted done.

At this point, she was pretty harmless. She was cornered, but tired. The twenty-man police force could easily have held her there indefinitely while somebody called a veterinarian with a stun dart. That would've rendered her unable to run and, once things calmed down, we could have reloaded her on the trailer and brought her home to settle for a few days before going back. But no, that would have made too much sense. Way too much testosterone had assembled in front of that cow, with way too much firepower, to do something as simple as sedation and gentle handling.

The firing squad began. Not one shot, nor two. No one has actually ever been able to piece together exactly how many rounds it took. But the consensus of all involved puts the number at somewhere between forty and fifty bullets to finally bring down the poor beast. Of course, since she had set foot in a federally-inspected slaughterhouse, the meat couldn't be given to the folks in a neighboring ethnic community, which would have appreciated the bonanza. At this point, the meat would have been no different than the venison from a hunted deer. The whole carcass had to be taken back to the slaughterhouse and disposed of — just thrown away.

An extremely dejected — and tired — intern finally made it home. Loss? Only a couple thousand dollars. Lesson? Priceless.

- Lesson 1: Park flush to the corral gate.
- Lesson 2: Cows are different than young steers.
- Lesson 3: Don't expect reasonable behavior out of men wearing uniforms and toting guns.

Polyface wasn't the only one getting schooled that day. The next day, the Harrisonburg Police Department called Tommy and asked to come over to his slaughterhouse so he could show them, on a cow skull, the proper placement for a fatal shot. Good grief.

Come to think of it, if we strung all these woeful tales together (and didn't forget the uplifting ones), we could indeed write a whole book. Maybe *Legends of Polyface* would be a good title after all. I'll have to think about that one. Let's draw this to a close. I hope by now even the most rabid believer in free intern labor can see that it's not free. An educational setting develops amidst a backdrop of, shall we say, interesting occurrences. Education is expensive. And at Polyface, we don't even charge for it.

Chapter 8

Spirit

A successful internship program requires both the mentor and the intern to put something in to it and draw something from it. You can look at it like a huge balance sheet. The problem is that the figures are subjective. If either party feels like they're putting in more than they are getting out, resentment develops. That's not a good arrangement. I've spent a good deal of time in the previous chapters developing the spirit and attitude mentors need to create a good internship environment. Now I'm talking to the would-be intern.

I liken internships to immersion experiences. It's like baptism. You're completely overwhelmed. You've got a new living situation, new work situation, new learning situation, new food situation, new entertainment situation, new financial situation, new value situation. That's a lot of newness to absorb all at once. On the human stress curve, this is an off-the-charts spike. I'm sure this accounts for the routine two-to-three day sicknesses encountered within the first two-to-three weeks of an internship. It's a lot to take in.

All of us like our routines. We like to get up at a certain time, go to bed at a certain time, think a certain way, eat a certain way, and do certain projects. Every intern arrives with assumptions.

Some interns think we're going to spend the day out sitting in the pasture, chewing on grass stems, letting the cows sniff our faces, and discuss the world's great problems. Some interns think they'll be able to go partying in town at night and sleep in for an extra hour the next morning.

Some interns think our family believes dating is wrong, contraception is of Satan, and watching Sunday afternoon football at Grandma's house is taboo. Others think we never eat chocolate or potato chips or heat water in the microwave. Some think we're evolutionists and vote Democratic. Others think we're quasi-anarchists who think every government agent is a demon. Others think our renegade ways border on rebellion and are therefore either anti-Biblical or un-American.

I won't belabor this, but you get the picture. Our farm and family have been put into every box imaginable. So here's the first thing necessary to get off on the right foot with your farm host: accept everything. Forget the judgment up front. Reserve that for later. I never apologize for being judgmental, but judgment prior to information is prejudice. Give it time.

That's why I like books. Our culture has devolved into sound bites, with almost everything presented as a "Tweet." It's absurd. TV news shows limit topics to two minutes of discussion. Shorter attention spans require sensationalism and stereotypes in speeches. When you look at the transcriptions of public discourse like the Lincoln-Douglas debates, you realize no audience alive in modern America would abide such thought-provoking speechifying.

Yet, given enough time to develop thoughts, most of us can find a nugget of truth in the perspective of even our most enigmatic foe. When I visit a farm, I don't make any assumptions about it. Instead, I try to assume a spirit of humility, openness, and acceptance.

I haven't walked in that farmer's shoes. I haven't lost a child or been through divorce or suffered sexual molestation as a child. Many people who look askance at my advocacy for property rights and libertarian tendencies don't appreciate what our family went through when we lost our farm in Venezuela in 1960. Those experiences color your life, your worldview, your politics.

One of the farm families I respect most has three autistic children. The wife and mother was a professional ballerina and today believes that the demanding dance and diet regimen created significant physical deficiencies in her body that were then expressed through these children. That's an incredible adjunct to their farm responsibilities, and in many ways, it comes first. These kinds of situations color how a family interacts with interns and how they set up their farm, their free time, their entertainment, their finances.

The point is that this incredibly intimate on-farm experience offers the intern a window into a given family's situation at a point in time. The successful intern will put her judgments on hold for awhile, and see the bigger picture of which she is just a passing part. Let the relationship develop. Ask caring questions, not judgmental questions. If you've been somewhere and seen someone do a particular job differently, don't carp to the mentor that, "You're doing this wrong. I saw it done this way over at XYZ farm and it was much better."

Chances are there's more to the procedure than you know. Devote yourself to accepting the protocols and techniques of the mentor you're with. Master them. I've learned something from every single farm I've visited. Sometimes it's simply a slick gate latch. Submission and acceptance will open your mind to the most learning possible. Don't worry about the why of everything; just go with the flow, follow instructions, and do it all happily. No grousing, no complaining, no whining, no pickiness.

Jump in. It's all about immersion. To keep the analogy alive, don't circle the pool like a finicky child, dipping your toe in to see if the water is too cold. If you've been picked as an intern, all the circling was finished when you said, "Yes!" Now it's time to jump in, full body, completely committed. You'll make a bigger splash that way, and the mentor will love you for it.

Some interns exude success. When we do our two-day checkouts, coordinating trains, flights, in-coming and out-going interns, it's a massive logistical process. One year we somehow missed being at the train station to pick up one of our applicants. Rather than sit down and call, he started walking to the farm, through Staunton, ten miles out into the country. We were all out in the yard having a picnic when we saw this guy walking down the lane carrying a backpack.

"Who are you?" we wondered. And though he'd just spent

two hours walking long country roads from the train station to the farm, he looked determined — and happy — to be there. As you can imagine, we all looked at each other and decided on the spot, "You're in." That's the spirit we're looking for. Just accept things the way they are and jump in. Ice on the water? Who cares? Algae and green scum on the water? Who cares?

One of my favorite memories of my dad, who was more innovative than I am if you can imagine that, was his pat answer when we kids would begin questioning him about some new gadget or plan. We were full of what-ifs. What if it doesn't work? What if the rope breaks? What if the tractor won't pull it? After a few of these, he would just grin mischievously and say, "I don't know, but we're going to know a lot more in a few minutes." What a great spirit of adventure. I'm not saying you check your mind at the gate; I'm saying engage your mind and your will to respond and observe at full capacity, meeting whatever you face with readiness. Be ready to receive information.

Many of our interns have never butchered an animal. At Polyface, every intern participates — substantially — in butchering animals. Our interns know this well in advance. It is a signature activity and core to our business. Any intern who wants to debate me about the ethics of slaughtering animals will not fit in very well. Such arguing also precludes the intern from learning very much. If an intern applies and then gets accepted into the program, the mentor assumes the intern doesn't have a moral or ethical dilemma about participating in the core curriculum of the internship. It's disrespectful--marketers call this bait and switch--to be accepted into an internship and then begin haranguing about the core activities you're asked to do. It's not fair to the mentor and it will make you a pain in the tail rather than a joy on the team.

I remember back before we opened up our internship opportunities to women, when I was out doing seminars, people would go crazy and give me standing ovations about our farming practices. But as soon as they found out we had not yet opened up internships to women, they stomped out in a huff. "Sexist pig," they muttered, under their breath.

What they didn't know was that we'd found out about two farmers who essentially lost their farms because female interns sued them for sexual abuse. I believed then and still do that these were opportunistic women who took advantage of a good faith relationship to line their pockets with a litigation settlement. Whatever the details of those cases, when we first began, we only had one internship housing option and a smaller staff which was already spread thin. Between our values as a farm and family, and the facilities and amenities we had on hand, expanding our program to both sexes wasn't practical or desirable at the time. It took awhile for us to process the demands — and risks — of creating a co-ed program and working out protocols to accommodate it. But that part of our process didn't slow down the tongue-wagging critics who just saw me as sexist.

Folks, I rub shoulders every day with people I consider nut cases because of the way they think. But I also learn from them. Lots of times, once I find out the background of their thinking, it makes sense. Doesn't mean I necessarily agree, but it makes sense and I can respect how they came to it. And that should be the mindset of an intern. That numerous women during that formative time prejudged me, tried to bully me, walked out of my presentation, called me a sexist chauvinist with no more information than that we didn't take female interns epitomizes callous disregard for assuming the best in the other person.

When we don't just accept things like they are, we shut ourselves off from the full learning environment. For some reason, it's much easier to find fault than agreement. We remember far more acutely what irritates us than what consoles us. That's why we call them stop lights and not go lights. We're inclined to focus on the stop and not the go. I don't know why; we're just hard wired that way. None of this means the intern can't ponder internally what makes sense and what doesn't on a personal level, or as a farming preference. But while you're learning, give your mentor the benefit of the doubt. After all, the mentor's the one in the business, and you're the one wanting to learn that business.

Remember, the mentor has gone to substantial emotional and financial risk to bring the intern into the picture. The right response from the intern is gratitude and acceptance. You have the rest of your life to do it your way and to find fault with the way the mentor did things. While you're at the mentor's table, under his roof, and handling his property and equity, you owe it to him and yourself to conform agreeably with his procedures and expectations.

You earn the right to disagree over time. Until you can do a procedure as competently and efficiently as the mentor, squelch the urge to make suggestions on how to do it better. We have a little saying, "Do it my way first." The thing you have to understand is that more often than not, a procedure has more background than might be obvious at first. Take time to develop proficiency in the mentor's procedure. Once that is mastered, then and only then can you legitimately question or disagree.

Don't be a prima donna about the different jobs that need to be done. At our farm, we preach that all jobs are sacred and none is more important than another. I know this goes against the grain of our stratified society, but ultimately if someone doesn't keep the toilet flushing, the CEO in his penthouse office on his million dollar salary won't have a very functional existence. I'm not suggesting that the two need identical pay, but I am suggesting that the two jobs are equally sacred.

Perhaps if the CEO had a self-imposed four-fold cap above janitor pay, the janitors would be more loyal. In dispensing the numerous jobs that have to be done around the establishment, none is inherently dishonorable or unnecessary. They are just part of the functioning of the business. They all have to be done. Somebody has to hold the sign to stop traffic while someone else runs the backhoe. One is not a good job and the other bad; both jobs need to be done in order for the whole project to be completed.

On the farm, the gamut can run from shoveling manure to riding along on product deliveries to helping to tote food and talking with customers. Most of us would rather ride along on a delivery and help with customers than shovel manure. That's fair enough. But the way you earn the privilege of the more agreeable job is to put in your time, volunteer, and show your willingness to happily do the less-agreeable job.

In a work environment as intimate as internships and mentoring, finicky attitudes toward job choices become apparent quickly. As a mentor, I can assure you that as soon as I see an intern consistently hold back from volunteering until all the jobs are on the white board, and then choosing the cleanest or easiest, I make a point to assign the dirtiest, least desirable tasks to that intern in the future. Every single task is a privilege, a valuable component of the entire process.

Shoveling manure drives fertility which drives production which enables us to have boxes of food to deliver to customers. I've written in previous books about what enabled me to quit outside employment and return to the farm full-time. I didn't have a nest egg. I didn't even have a business plan. My epiphany was realizing that I didn't consider any job beneath my dignity or intellectual capacity. I'm as happy washing dishes as pushing concrete in a wheelbarrow as selling beef to a restaurant. As soon as I realized how that spirit made me drop-dead employable, I made that leap into the risky world of full-time farming, knowing that everyone who employs people would love to employ me. That's real freedom.

That's the spirit for a great internship. Be the first to volunteer, even when you don't even know what the job is. Sometimes on our farm I tell the intern crew I need two volunteers, purposely not telling them what the job is. The ones who eagerly step forward, no holds barred, no questions about what it is, how long it is, how hard it is, where it is — those are the interns that quickly move up into more management and responsibility.

Mutual Profitability

Perhaps this is as good a time as any to broach the subject of a successful intern's number one priority: mentor profitability. While that may sound egregiously self-serving, it's the truth. Interns come full of expectations about the experiences *they'll* have — conversations, memories, and friendships. All of those hopes and dreams are noble and good. But what should consume the heart and mind of any intern is asking the fundamental question: "How can I help this operation be more profitable?"

I can feel the resistance coming from readers right now. The reaction might be, "What? Profitable? Why you money-grubbing, evil capitalist. Here you've talked about love and sacrifice and family and really all you're interested in is lining your pockets with free labor. A pox on you and your kind!"

You see, this is why I like books: because they give you

enough room to develop an idea — even a politically incorrect one — and recognize the naysayers. In defense of profitability, I hope that by now you understand that this is clearly not the *only* goal of an internship program. I've purposefully deferred introducing it to put it in the proper perspective.

But with all due respect, if an intern isn't thinking about and actively pursuing this notion of overall business and mentor profitability, a huge piece of the experience is missing. If we don't have a profitable farm, we don't stay in business. If we don't stay in business, we don't have an internship program. If we don't have an internship program, you don't have this opportunity to learn. Goodness, sounds like a Geico commercial where things deteriorate due to failure to buy their car insurance.

Business profitability is rightly demonized in modern culture because, too often, it *is* the sole consideration of a given business concern. In addition, the time span is usually too short. Profitability over a century is quite different than profitability over six months. The inordinate preoccupation with short-term gains found on Wall Street concerning profit has unfortunately undermined their credibility, relegating that whole investment model to the scrap heap in many peoples' minds. Yet profit is the lifeblood of business, and farming is a business just like any other.

Profit is what allows Polyface to invest in intern housing. It's what lets us do everything that we do. Recently I had a group of college students at the farm for a tour and at the end, several of them noted that I hadn't been keen on government solutions. Several of them asked if they didn't go into government jobs, where else could they go, or what else could they do, to be environmentally helpful? The notion that the solutions to the problems facing our culture could be solved more efficiently with private entrepreneurial business had never really entered their minds.

How do you think government gets the money to pay for the programs and people it employs? It comes from profitable businesses and individuals employed by them. Despite what some would have you believe, the money doesn't come from the Federal Reserve or stimulus packages, a euphemism for cranking up the printing presses and destroying value. It comes from hard working private sector economic activity, providing goods and services with personal capital, personal risk, and personal profits.

The most successful interns are the ones who ask, "How can I help you make money?" Obviously, making money isn't our only concern — it's one of several objectives that are all balanced with ecology, sustainability, charity, and integrity. But by putting the goal of profitability into the mix of an intern's aims, a carefulness and an awareness of the big picture arises that likely wouldn't otherwise exist. Such a spirit helps an intern think about wasting fuel, breaking or misplacing tools, losing gloves, breaking eggs.

Interns must be as careful about watching costs and stopping slippage as the mentor. Unless and until that personal ownership occurs, the intern works along with a misplaced notion of entitlement and importance. When interns figure out how to make their mentors money, guess what? They've just figured out how to be financially successful themselves. The past is prologue. The attention to frugality and financial stewardship in the internship will carry over into an intern's future endeavors.

The internship is an incubator for habit development. Habits are not just procedural how-to's. They include how we think, and how we behave. The privilege and efficacy of internship is that it puts the student proximate to a master who has earned his chops over time and trial. The mastery is functionally validated by the doing of the thing, in this case, farming. By that I mean that the master didn't attain the position by buying off, figuratively or literally, a board of inspectors who conferred a title or license. He earned it through doing.

I'm always amazed by the number of people who think successful people just got born with a silver spoon in their mouth. While that certainly is true for some, it is certainly not true for most. Even a cursory glance at wealth holding shows that most people considered successful simply worked harder and longer than most others. Those who think charity begins by forcing successful people to share via graduated tax percentages actually despise hard work, thrift, and all the attributes that enable someone to pay taxes in the first place.

Unfortunately, many interns grow up in this spirit that applauds demonizing successful people and come with a bit of a chip on their shoulder regarding mentor accomplishments. I know about one situation where the interns on a farm complained that the farmer didn't do any work. Folks, the farmer was 60 years old, extremely successful through innovation and hard work. Did they think he was in the house napping?

Of course he wasn't. He was finding supplies at the best quality and price. He was talking with a chef about buying pastured chickens. He was talking with an engineer friend about a possible design for a portable field shelter or clever seed planting device. Trust me, the mentor has stayed up more nights, put in longer hours, tried more things, and sweat more than the intern can imagine. Respect that; honor it. Good interns realize that mentors have done their time, so to speak, and need administrative time on the phone and at the desk to manage the operation that has now gotten significant enough to provide an intern opportunity.

When the intern approaches the whole arrangement with a driving spirit of helping the mentor to be successful, it creates a climate of living and habit that will establish similar success in the intern's future. I've already addressed the need for the mentor to have a similar desire for the intern. Focusing on the intern's commitment doesn't negate all the mentor's responsibilities and attitudes in preceding chapters. The fact that mentors arrive at their place of leadership by the habits and procedures they've honed competing in the real world offers interns a rich immersion opportunity.

In very real terms, profitability is the mentor's return on investment. When we choose interns, part of our discernment process involves asking, "Who can best leverage our investment toward both short-term and long-term profitability?" We aren't just looking at who can best perform tasks and team-building during the internship, but also who has the character and best chance to leverage our investment into their future lives. We're thinking about their chances for future success as they carry the "Polyface internship graduate" badge of honor. We don't take this lightly — it's our reputation as well as the intern's that's on the line.

We don't want to graduate a bunch of failures. What would that prove? By the same token, we certainly don't intend to graduate a bunch of robots. But we do know some principles that work. A bunch of them. Living simply — frugally — is all part of the plan. An intern who deviates from our lifestyle and these principles probably won't be successful. That's not hard-nosed; it's just wisdom. Interns who realize that their success is wrapped up in assuring *our* farm success are far more apt to learn and adopt these principles.

Everything we do has a reason. Too often interns, especially farm interns, think many tasks are brainless. Successful interns acutely observe everything the mentor does. Mentors don't waste motion; they don't squander effort. Every single thing, from shoveling dirt to weeding green beans, has deep thought attached to it. Eliot Coleman, guru of extended season commercial gardening, exudes this in all his writing and garden techniques. I've had the pleasure of spending a little time with him, and every time I come away bursting with admiration for the extent to which his brain is engaged during every seemingly mundane task.

Good mentors have gotten where they've gotten due to a lifetime of accumulating skills and techniques. I visited a farmer recently who complained about moving chicken shelters. He was on steep terrain. His house and outbuildings were located near the base of the hill, so he naturally started his broilers there and proceeded to pull the shelters up the hill. I suggested that he take the shelters to the top of the hill and pull them down hill. It revolutionized the operation. He'd never thought of it.

How could I come up with a solution so quickly? It seemed obvious to me. The reason is because I've pulled shelters for many, many years. Over time, I've learned how much easier it is to pull them downhill than uphill. Usually a person figures out how to do things easier or better over time. The whole idea of internship is to compress the mentor's lifetime of development into an intensive, intimate, informational and technical transfer situation so that the next generation doesn't have to start at square one. The way this is supposed to work is that the intern starts where I finish and goes on with yet another lifetime of refinement and betterment.

If an intern thinks any activity is mindless, she'll miss important details that can make or break future endeavors. An intern should assume that every single thing a mentor does is for a reason. One of the most common mistakes at Polyface, for example, is wrapping wire at the end of a fence. Over and over I stress the ninety-degree wrap — in other words, take the end and wrap it around perpendicular to the wire coming to the end. A tight wrap won't slip. But invariably, interns wrap on an oblique and it will slip sure as anything. This is an example of how every procedure contains information and experience.

The level of this embedded knowledge can't be immediately appreciated by the casual observer. That's why interns must not be casual observers. When you're out working with the mentor, get up where you can see. What really is the technique? How does he hold the hoe? How does he grasp the weed — front end or back end? Which fingers? Both hands or just one? If only one, what's the other hand doing? Foot placement? Body placement? Center of gravity? Eyes in relation to hands? Are you with me? Every single thing; every single thing; every single thing has a host of nuances. Successful interns assume this, which drives their observation and mimicry.

Teamwork

Good interns exude a spirit of teamwork and fellowship. They aren't loners, but in the spirit of community jump into the full range of social, experiential, and educational opportunities. This means you carry your weight. It means you show up on time. It means you honor your responsibilities. When it's your turn to wash dishes, you don't have to be reminded.

Good interns aren't hermits. They don't have to be party animals by any means, but they do need to enjoy the people around them. If you don't enjoy your team, what's the problem? Look in the mirror first. Being a team player means that if I'm not fitting in, my first judgment is toward myself rather than the other folks. If they all seem happy and appear to be having a good time, what's my problem? Looking inside before looking outside is the mark of maturity and personal growth.

This isn't to say you shouldn't have your own personal projects. Some interns want a personal garden. Some want to do woodworking. We've had some who wanted to tan leather. We had one who wanted to eat as many different animals as possible: raccoon, skunk, possum. One of our interns came to the castration lesson with skillet in hand and cooked the mountain oysters for all to enjoy.

Did you catch that last phrase? "For all to enjoy." All of these individual and personal interests are great as long as they aren't exclusive. Most mentors enjoy watching the various personal projects some interns want to undertake, but if others are excluded from the process, these projects destroy the team rather than build it up. Good leadership involves bringing others along with you.

Desire to Work Hard

Finally, good interns actually want to work hard. I mean work hard. If you want to endear yourself to a mentor, be the first one out after breakfast. When the mentor says he'll meet you at five in the morning for some special project, make sure you're waiting on him, not the other way around. Everyone else dismissed for the evening? You be the one to ask, "May I help you do anything else before dark?"

Go the extra mile. Ultimately each intern gets out of the process what he's willing to put into it. This is the principle of sowing and reaping. Sow sparingly, reap sparingly. Sow abundantly, reap abundantly. You can't out give the experience. Give it more than you think you've got and the investment will yield more return than you ever could have imagined.

If I'm sweating enough that my pants are wet clear down my thighs and I look over to the intern helping me and she hasn't even broken a sweat, what does that say about effort? Ultimately, only you, the intern, can determine how much you'll get out of the internship. In fact, it's your responsibility. If you want to go to town and drink after dinner rather than hanging around to see what else can be done or to just be available, you'll get less out of your internship experience.

But if you devote your undivided attention to this experience, you'll get more. When the cows get out and you have the distinct privilege of being able to help get them back in, you'll learn far more than the interns who went to town to party that evening. It's in the crisis, the crucible of stress, that the real make-it-or-break-it principles shine forth. Being available and ready to participate in those off-times, to truly own the farm, will enrich the experience.

Spirit

You know Murphy's law that everything bad happens at the most inopportune moment. That's true on a practical day-to-day level. The chances of the cows getting out are in direct proportion to your anticipation of not doing anything with cows at that time. The corollaries are often funny. The propensity of the bread to fall jelly side down is in direct relation to how long it's been since you cleaned the floor.

I wish I had a nickel for every time I've seen the tail lights of the last intern car leave the farm for some town hang out time just as I notice the heifer we've been watching suddenly decide to calve. Murphy's corollary: the chance she'll need assistance is in direct proportion to the difficulty in finding anyone to help with the delivery. I'm not suggesting that interns tie themselves with a chain to the farm, and I certainly don't think having a good time is wrong. But we've had some interns who want to go to town (and you can imagine what that means) several nights a week. Others are content for once every two weeks. Guess who gets involved with those after-hours deep learning experiences?

Every year, within just a couple of days, the team leaders emerge. As we staffers compare notes, our special interns rise to the top. Almost immediately the season's die is cast: we can trust these ones with anything; better wait awhile on those other guys. Your goal as an intern is to be the cream. No excuses. No victimhood. Just an exuberant, happy, satisfied spirit. You show that spirit in the internship and life will rise to meet you. Now go have a life-changing internship.

Fields of Farmers

How It Works

Chapter 9

Vetting and Selecting

E very farm internship program reflects specific nuances, customized refinements that speak to individualized resources and people preferences. No two programs will be identical because no two farmers are identical. In this book, I've attempted to address issues without listing all the possible solutions. I've opted, instead, to explain the whys and wherefores of our Polyface farm program, in hopes that others will take what works for them and adapt it to their aims

We've been in this intern business now for a couple of decades and have learned a few things. We have plenty more to learn, but we're much smarter about this than when we started.

We've been through the, "I have this child that I know would benefit from being at your farm for the summer" fiasco to the relationship complexities of extremely conservative Christian homeschoolers away from home for the first time living with tree-hugging, liberal, free-love collegians shacking up with short-term partners. Whew! Long sentence; longer discussions, let me tell you.

In some senses we're boss, dorm supervisor, campus chaplain, coach, security guard, ombudsman, and more all wrapped into one. These are a lot of hats to wear. Each one takes some adjustment and some certainly wear easier than others.

How It Works

I'll lay out as specifically as possible the nuts and bolts of our Polyface internship program. Each year we accept applications from August 1st through August 14th. That is significant. When we first started, we accepted applications any time of the year, but as things progressed, we needed to formalize the process and the dates. In general, even fledgling programs should create an air of formality because it's the mark of professionalism.

While some may view this as unnecessarily rigid, I'd argue that a bit of rigidity creates clarity. The more an internship program is run spontaneously, kind of like a half-baked idea, the more likely it will degenerate into confusion and misunderstandings. Don't be afraid to formalize it, to structure it from the get-go.

Such rigidity also weeds out the folks who are too free-spirited. Think about it: Do you really want a free spirit that thinks you're nuts when you scream, "Get out of the way!" If you run over a free spirit with the tractor, you might free the spirit from the body, but that won't be a liberation you'll want on your farm.

Look at your farm work flow and plan accordingly. In the early years, we got into trouble sometimes by accepting people too quickly. Obviously you can't expect to complete the vetting and choosing process too fast, but you also don't want to have too long an engagement, so to speak. People can change their minds. We've found that a six month process seems about the right amount of time from acceptance to arrival for the internship. Much longer than that and things change. Shorter than that, and applicants feel cramped, like they can't adjust their lives.

Remember, for many potential interns, this is a major life change. For those who are still in college, it usually means taking a semester off. For people out of college who are already working somewhere, it means procuring a leave of absence. That can be a lengthy process in a large company. Pets, apartments, and spouses need to be considered. It takes six months to do all that.

We use our website to publicize internship opportunities. It is up 365 days a year and anyone in the world can go to *polyfacefarms.com* to see the sales pitch and description. The website includes the point that we are "very, very, very discriminating." This is a tongue-in-cheek way to explain that we choose who we want to. We don't believe it's necessary to defend our decisions. We could use the word selective rather than the more militant discriminating, but we've found that strong language helps weed through the folks we don't want. Anyone who can't appreciate a little humorous jab at prejudicial hiring and acceptance requirements promulgated by the government probably won't fit into our rough-and-tumble freedom-loving table conversations.

Polyface is our business. Our farm. We don't ask for government subsidies, concessions, or handouts. We can accept or reject anyone we want and we honor the right of any other farm to do the same. It's a not-so-subtle way to categorically reject political correctness, affirmative action, and all the other diversity-based anti-talent protocols out there dominating institutional selection processes. It does indeed matter what you look like, how you handle yourself, whether you can carry two 5-gallon buckets of water, and not wear entitlement on your sleeve. This has nothing to do with race or ethnicity; it has everything to do with demeanor and hygiene.

I routinely run into people with dread locks asking if they comply with our requirement for an "All American boy and girl look." The answer is no. Folks, please, we're in the clean food business. Our customers aren't interested in being handed a package of artisanal meat with hair that by definition can't be washed lest it mold. That doesn't mean I can't love people with dread locks. Goodness, I love sex offenders, but I don't ask them to babysit my kids.

If you want to make a statement with piercings, tattoos, and dread locks, that's fine, and we can have a great conversation. But your statement won't drive my delivery truck or represent me at the sales counter. Shock value is fine — on your time. Not on mine. The only real endorsement that matters to us is the expectations of our customers and the fit with our own farm standards. We set our intern standards as high as the standards we set for our farm. One of the reasons our farm has been successful is that we take great pains to accept team members that fit.

I'm sure some people reading this right now are ready to throw this book away. Perhaps it's time to grow up into civility and respect. Just because we don't take some people does not mean we disrespect them. Goodness, look at the requirements for being a Rockette in the New York performing dance company. Talk about tight stipulations. But without that tightness, the dance wouldn't look nearly as amazing as it does. In all fairness, our tight policies may occasionally keep us from choosing a great intern. I don't doubt that at all. But on the other hand, I find it interesting that I'm accused of disrespect for having a tight selection policy but the young person who wants to come is not accused of disrespecting me for wanting to look or act a certain way. What's good for the goose is good for the gander here.

That doesn't mean other farms won't have different standards. Some farms specifically select juvenile offenders, or people with a prison record, because their mission is to use the farm as a fresh start catalyst for troubled people. I applaud that —it's wonderful. I categorically defend the right of those farms to make their own choices, including limiting internship candidates to those people in need of rehabilitation.

Our farm has a different ministry. We're showcasing and innovating agricultural prototypes that are environmentally, economically, and emotionally enhancing and facilitating their duplication throughout the world. If we decided to become a reform farm, that would distract us from our primary mission. Perhaps we could look at it this way: by freeing us up from the reform mission, we're able to develop the prototypes that a reform farm can quickly and efficiently copy to free it up to minister to troubled people. Everyone doesn't share the same mission. That doesn't make one right or wrong; it creates service diversity, and that's a good thing.

On the website we articulate the attitude and spirit we want. We seek bright-eyed, bushy-tailed self-starters. We don't want hangers-on. We don't want people waiting for a hand-out or sporting a victim mentality. Dependency has no place at Polyface. Being up front like this ensures that we get the best of the best. For too long, farming has suffered from a cultural stereotype depicting backward, peasant-styled personas. We're promoting the Jeffersonian intellectual agrarian. The farmer-entrepreneur. Perhaps we should call her the agraria-preneur. Eager to learn, eager to work, eager to stretch and be challenged.

Unlike most businesses, government agencies, and colleges,

we don't even ask, anywhere, for religious or ethnic affiliation. To me, that is far more open and embracing than an application process that does ask for those things. If we're really going to be fair and open in our process, why even ask for race or ethnicity or religious affiliation? I find it fascinating that government paperwork routinely has a non-discrimination standard at the bottom of the page, but in the form still demands that kind of information.

Here is what we do demand, though. We demand that interns respect and honor our values while they are here. Our family and farm have specific and distinctive values. For example, anyone familiar with me knows I lean heavily libertarian in political thought. But we don't ask anywhere, or at anytime, if our interns lean libertarian. However, we don't accept interns who covertly or overtly undermine this position. It's certainly not a taboo subject and we don't mind discussing liberal versus libertarian policy. But we don't tolerate agitation and we definitely will not stand for arguing in front of customers or undermining our values to customers. We present a unified, content, happy team to visitors and customers. If you can't bury your agenda for four months, we don't want you.

Our family, as is no secret to most, takes a strong, literal, creationist stand. We believe the Genesis record, believe that science corroborates a one-week creation explanation far better than conventional evolution, and often use the term creation to describe the physical universe. We don't ask, and really don't care, if our interns believe this or not. Honestly, it doesn't matter to us. We aren't running a Christian camp retreat center. What does matter is that the interns respect our position. We're glad to discuss it. But we can't abide snickering and undermining whether behind our backs or in public. You can switch that dial off for four months. If you can't, we don't want you.

I'm sure there are plenty of evolution-agreeing farms out there. As a mentor, don't be afraid to define, clearly, what fits for you. And as an intern, be willing to realize how intimate and costly, both in energy and emotion, this farm openness can be. The farm is risking its reputation on you. It's not fair to undermine the brand, the track record, the reputation of the farm. The farm persona represents the sum and substance of lifetimes. Respect that. Be content to find the arrangement that fits. It's the honorable thing to do.

Every one of us makes judgments every day about what fits for us. Our circle of friends. Our spouses. The jobs we choose. Professionals we employ — doctors, small engine mechanics, plumbers, electricians, roofers. You name it. We're constantly judging based on fit. We like to surround ourselves with people that fit. Goodness knows we're crammed together with people who don't fit often enough. When it comes to the most intimate part of our lives, we inherently and necessarily seek out what fits.

It's like our first apprentice Tai told me, "I like ponds, and I like trees. But trees don't grow in ponds." I thought that was a good way to look at it. I enjoy diversity, but too much in the same place creates discord and war. Our farm is a sanctuary, not a war zone. Our boot camp is where we establish unity so we can fight the wars that are really worth fighting. We certainly don't ask interns to agree with our values or our position, but we do demand attitudes and spirits that will respect our positions so we can all watch each others' backs during the day.

I know this sounds offensive to some people. I'm well aware of the attacks that vilify our family and me specifically for daring to describe the appearance and attitude that fits for us. Guess what? It's our farm. We're the ones who make the calls and knock on doors to get customers. We're the ones who stay up with sick calves. We're the ones who spend sleepless nights keeping the coyotes out of the chickens. We're the ones who fix the fences when storms knock trees over on them. We're the ones who carry water, gut the chickens, move the pigs, build the intern housing, pay the bills, and face bureaucratic wrath for failing to dot an "i" or cross a "t" on a host of inane government forms.

To those offended by this discussion, I simply offer you the hand of a farmer and invite you to wade on in, the water is fine. Join me with your own individualized swim stroke. Have at it. We'll be friends anyway. The world is plenty big enough to accommodate your way and my way. I'll respect yours and ask for your respect in return. We don't have to agree. I do believe that our Polyface stance and demeanor have been foundational to our farming success. When our position fails to create success, I hope and pray I'll be humble enough to change the position. Until then, I'm incorrigible. Ha!

Applying to Polyface

In addition to a brief description of the program and the kind of person we're looking for, our website makes it very clear that we only accept applications from August 1st through 14th. Period. No exceptions.

Every year we have latecomers bemoaning that, "I didn't know about this until now!" They beg for leniency. Too bad. Sorry, you'll have to wait until next year. Deadlines create efficiency and functionality. And deadlines create a professional decorum, subliminally saying that, "We've thought about this, created a schedule, and invested planning into this program." A haphazard, catch-as-catch-can approach will not set a bar high enough to stimulate jumping. You want people jumping for the bar.

Yet the same application window applies for the too-eager types. Routinely we receive an email or letter saying, "I know your website says applications only accepted August 1st through 14th, but I'm trying to get a leg up on the next round by getting my name in front of you now in June so you'll remember me when I put my name in the pot in August." Folks, that's a sure way to *not* be chosen. What does such an attitude say about our boundaries? Anyone who sends a letter like that will be the one pushing the envelope on respect and appreciation while they're here.

Our application process starts by simply putting your name in the pot. We don't require resumes, references, or affidavits. We do pretty much require an email so we can communicate efficiently from that day forward.

When we receive the request between August 1st through 14th, we reply with a questionnaire. I won't share every single question we ask. Please understand that I'm sharing the heart and soul of our business here, creating vulnerability on an almost unprecedented scale. So I'm holding back a little bit — kind of like not showing you my underwear. You can look in my clothes closet, but not my underwear drawer. Fair enough?

The first question asks applicants to describe Polyface in their own words. We want to be sure that they understand what we're about, to demonstrate some knowledge about this farm. While this may seem unnecessarily simple, you might be surprised what surfaces in spirit and expectation from this question.

Then we ask why they want to come here. This drills a little deeper into perceptions and often includes their life goals or objectives. We want people with depth and purpose, not looking for a thrill or some cool experience. We try to separate the whimsical from the thoughtful, the impulsive from the well-considered.

We ask what they've been doing the last couple of years. This gives us some background into abilities and natural inclinations. Often special skills surface here, like welding, woodworking, mechanics, or teaching. We've had everything from professional dancers on cruise lines to Hollywood film editors to Alaskan fishermen. Every year when I read the answers, I'm overwhelmed by how eclectic the field is. Profitable integrity farming appeals to the broadest spectrum of people imaginable. We have the distinct privilege of combing through this sea of applicants to find the ones who will synergize most harmoniously with our farm. What a privilege.

Each farm needs to customize such questions to fit its operation. For example, unlike many farms, at Polyface we need to ask, "How do you feel about killing animals?" All of our interns must process chickens, turkeys, and rabbits. At some point in the season, we usually farm-process a pig as well. One of our field trips is usually visiting the slaughterhouse to see our steers killed and processed. We need to know, straight up, if we're in for a queasy stomach or not. We've never had a vegan or vegetarian intern. We have no problem with vegans or vegetarians, but they don't fit with our mob-stocking herbivorous solar-conversion lignified carbon-sequestration fertilization ministry.

For produce farms, I suggest asking straight up, "Do you mind picking beans or weeding carrots for four hours a day?" Here, you're trying to communicate the most potentially distasteful aspect of the internship to allay any ishy-gishy, prima-donna attitudes. We want no hesitation. We don't want reluctant attitudes; we want embracing attitudes. No passing judgment; just git 'er done.

We also ask if there is anything at all that would interfere with fulfilling the obligations of the internship in its entirety. Again, this is self-protective. Our internship program starts June 1st and runs through September 30th. Many times an applicant wants to start back to school before September 30th, or can't get to the farm before June 1st. In reality, our new crop of interns must arrive a couple of days prior to June 1st so they're ready to go at daybreak on June 1st. Again, it may sound harsh, but we give no exceptions. The definite start date enables us to go through orientation only one time.

Part of the reason we're so hard-nosed on these things is because too many interns see this as a cool experience rather than the template that will help format their future. I'll discuss this a lot more in the chapter about intern expectations, but if the farm begins making exceptions for starting and ending times, among other things, it simply waters down the entire process into an extended theme-park stay instead of the real nitty-gritty farm experience we guarantee. By signing off on the commitment, interns are forced to contemplate, seriously, the ramifications of this four month investment. It gets rid of the whiners right off the top.

Our standard answer when the whining starts is to just smile and advise them to try again next year. We might add that, "Obviously the timing is off for you, but if the attitude is right, you'll be back and we'll take another look when it does suit." That keeps you from getting into an argument and maintains the line of authority. If the applicant continues to whine about an exception, then it confirms that their attitude is a problem. If the concession-requesting applicant, on the other hand, takes the "next time" answer in stride, smiles, and complies, then you know you've found a pleasant attitude.

We take very seriously the responsibility created by the internship program as well as the chain of command that makes it function efficiently. If you start trying to cater to individual wants and desires, the whole program will collapse. This is not a democracy. It's not even a republic. It's an enlightened despotism. A benevolent royalty. Not a tyrant, but a loving, responsible guide who knows more about what it takes to be successful than the intern does. So shut up and listen. It's called leadership. Visionary leadership. Believe it or not, all of us yearn to follow great leaders.

Now it's August 14th and the last applicants have put their names in the pot. We've sent out the questionnaire, which must be completed and back in by September 1st. Just for the record, early birds don't get special consideration, nor do the ones who slip in under the deadline get penalized in our thinking. This year, two of the interns who ultimately moved up into our apprenticeship program actually sent in their questionnaires within minutes of the midnight deadline. Why? They wrestled, up until the last minute, about whether this was really something they wanted to do. We understand that and don't assume that the folks who get their answer in right at deadline are procrastinators. This isn't a research paper. It's a big life decision, and not one to be entered into unadvisedly or hurriedly.

Numerous initial inquirers ultimately don't send in the questionnaire. They get scared off by the questions or upon further reflection, decide to do something else. Last fall one-third of our initial names-in-the-pot didn't fill out the questionnaire. I expect the question about killing animals scared off the lion's share (intentional metaphor for special effect). That's fine. Plenty of non-animal farms exist. Going to a carrot-killing farm will be a much better fit for all concerned.

Once the questionnaires come in, on September 1st, Sheri, who currently handles all of the e-mails, then creates an evaluation spreadsheet. Since we have different housing arrangements for male and female interns, we separate them in the evaluation process. Historically, we've taken more males than females because that's how our housing is set up. We haven't found one or the other to be superior. And we're absolutely open to flipping that ratio. Every year is different and we make our calls based on that year's talent pool.

Sheri alphabetizes the names on the evaluation spreadsheet. Now the real work begins. Teresa and I, and Daniel and Sheri, read every single questionnaire. In 2012 we had 203 returned questionnaires for eight internship spots. That, dear people, is one of the reasons I'm writing this book. We've seen the interest in farm internships skyrocket. It's not just our farm; it's everywhere. As far as I can tell, the shortfall is not in intern interest; it's in mentors. I hope this book will remedy some of that. My heart breaks for all those excellent applicants who we can't take into the Polyface program. Sure, we wouldn't want many of them, but that doesn't mean their hearts aren't in the right place and they wouldn't be perfect for someone else.

It's disconcerting that it's easier to get into Harvard and

Yale than to become a Polyface intern. The tragedy is that we take less than four percent of the applicants. The good news is that this percentage allows us to get the cream of the crop. Nothing pleases me more than to see smart, sharp, articulate young people attracted to land stewardship in farming's hands-on style. What a contrast to the ones who go straight from academia into government regulatory bureaucracy, perhaps in some agriculture or environmental agency, and spend the rest of their lives telling farmers what to do. They become arbiters from the computer screen only, with no practical on-farm experience. The day of the rock star farmer is at hand. Farming can absolutely return a white collar salary. The farmers most apt to find that reward are the ones who combine brain with brawn, utilizing strategy, insight, and innovation while working at things until they're physically done.

On September 1st we also begin reading the questionnaires. It's equivalent to reading *Moby Dick*, except, as Daniel says, it's even worse because it's the same chapter over and over. We split up the questionnaires among the four of us and spend the next two weeks soldiering through them. The evaluation spreadsheets have a line for comments to help jog our minds when discussion time comes. The spreadsheet also contains a column for yes or no. I like to grade them A, B, C, D or F. Teresa likes to put Y- or Y+ or N- and N+. The important thing is to pass judgment.

Many people ask how in the world we can separate the applicants based on just a few simple questions. Like anything else, it's an acquired skill. When we first began this process, discerning the yesses and noes was much more difficult. But over time, you begin spotting words and phrases that either endear or defeat. Sometimes the red flag is obvious enough to not even have to finish reading the questionnaire, but that's rare. Most of them, however, are far more subtle and take a careful read, then a re-read. We desperately don't want to miss a good candidate or a bad one. Answer length doesn't matter to us. We're not looking for dissertations. Longer answers don't increase acceptances; shorter answers aren't a liability. We've never wanted the process to favor proficient writers. Punchy, powerful phrases are just as good as wordy answers. We don't give credit by the word. We're looking for directness, thoughtfulness, values, and a good fit.

Over the years, we've modified the questions, deleted some, and added others. I'm sure we'll continue to do that as we hone this significant part of the process. Because the relationship is intimate, the vetting process is the make-it-or-break-it aspect. It determines if we're going to have a good year or bad year. One bad egg can cost us financially and emotionally more than most folks can imagine. Poor help is far worse than no help at all.

Once we've all independently read the questionnaires, we get together around the kitchen table and go through the list, making five piles of common denominators around our assessments. The first pile is four yesses, second pile three yesses and one no, third is two and two, fourth is one yes and three no, and the fifth, all noes. Believe it or not, at the end of reading the 203 applications in 2012, we had 40 all yesses. We also had 40 all noes. Remember what I said earlier about when the family doesn't agree on things, it's best not to start this process? The fact that all four of us have this depth of agreement is crucial. In short, we know what we're looking for.

This initial selection process is for the candidates we invite to the farm for the two-day checkout. They must come to the farm, eat with us, work with us, and sleep on the farm. Because we're going to invite a few extra candidates to our two-day, on-farm, check-out visit, we need more than the forty unanimous yesses, who are automatically in. Obviously the next pile is the three yesses and one no. What ensues is a diplomatic discussion wherein the one "No" voter defends that designation to the other three. Sometimes the no wins and other times the no acquiesces to the wish of the majority. At no time is any one coerced into a change of mind. Each one of us has complete veto power.

Finally, we open up the negotiation for anyone to bring to the table someone who hasn't been chosen, but whose questionnaire indicated a superb candidate that somehow got overlooked or misjudged. That allows even a three-noes candidate to be resurrected for examination.

We've been using this process for years, and it's never a fight, but an enjoyable give-and-take that actually helps us as a family to appreciate the feelings of each other. I find this experience enlightening and a coming-together time for all of us. We usually sit around drinking raw milk and eating fresh-baked brownies or chocolate chip cookies to bring a festive air to the serious business of evaluating. After slogging through all these questionnaires for two weeks, spending every spare minute thinking about them, the family negotiating time is like payday. We've arrived at this point and can enjoy the fruits of our labors. It's not a dreaded time, but a delightful anticipation of next year.

Remember, all this happens around September 15th, which is right at the end of the current internship season. If we made some mistakes last year and have suffered through a less-than ideal summer internship situation, we look forward to this new crop with optimistic anticipation. On the other hand, if we've had a stellar summer with the best group imaginable, which is normal, we look forward to this new crop eagerly, anticipating another fantastic group of young people to love and encourage. Either way, the session is delightful and not arduous.

The time may come when we only take the unanimous yes pile. That would simplify the process a little bit. It all depends on the numbers.

As soon as we make our selections, Sheri sends out congratulatory e-mails to the successful candidates and Dear Johns to the others. As part of their congratulatory e-mail, the candidates are invited to the farm for the two-day check-out visit. This is the next big step in the vetting process. We offer a two-week period for this check-out opportunity. Candidates may pick any two days within that period to come and work with us for 48 hours. No more and no less. No exceptions. They cannot stay for three days and cannot slip it in for one day. It's two days.

If the time doesn't suit, too bad. It's not fair to the other candidates for someone to come at a different time, not having to rub shoulders with the others, receiving special one-on-one attention and not having to be part of a team. Besides, this check-out visit is disruptive enough to our lives without spreading it out to other times. One visitor is as disruptive as five. I'm not complaining about the disruption — make no mistake, we love being disrupted for this selection process. It comes with the territory. The more we disrupt ourselves during this time, the better the process works. We could call it investment disruption.

We believe very much in the confluence of forces. If it's

meant to be, you'll work it out. Candidates who fight us over time or check-out protocols won't be accepted. All of this toe-the-line, no-exception policy separates the dedicated, passionate ones from the posers. Posers normally want to pick their jobs, pick their times, pick their food. Team players jump right into the fray, tumble around without complaining, get with the program, doing it all happily and joyfully. We believe strongly that a person makes his own happiness. We've had young people apply from extremely difficult circumstances but move heaven and earth to comply with the stipulations. They receive rave reviews.

Of those invitees, normally half a dozen won't come. Either our check-out dates don't work or they have a change of mind or whatever. What we really want are about four or five times more checkouts than the number of interns we need. This may seem excessive, but again, we'd rather be overly inclusive on the front-end so we can be more selective on the back-end. Meeting more candidates never hurts. The more you meet, the more options you have. We like about four or five times the number we'll eventually select. In other words, for eight spots we want 35-40 check-out candidates.

We've tried many different approaches for the two check-out weeks. We've split these two weeks from one week in early December to one week in early January. We've put them both in early January. We've put them both in early December. In general, we like the two weeks back-to-back with no break, primarily because the logistics to get the first one arriving and the last one gone are harder to do with a break in between. Better to just keep things rolling once you're into it.

As you can imagine, this two-day check-out is a major undertaking. We house, feed, and work a constantly fluctuating group totaling 40 young people over the two-week period. Running to the airport (candidates must handle their own transportation to the area though we're glad to pick them up at the airport or train station) sometimes three times in one day, coordinating beds and meals, and coming up with lots of grunt jobs for this many people is grueling. If we knew an easier way, we'd sure jump on it. As far as I know, nothing compares to actually living, eating, and working together for a couple of days in order to find the ones that fit.

In order to put each of these potential interns into as many different situations as possible we rotate breakfast, lunch, and dinner between Daniel and Sheri's house and Teresa and my house. Obviously at Daniel and Sheri's we're able to observe how the candidates interact with the children and at our house we see how they deal with old fogies. Both of these interactions are important in this intimate experience.

In addition to Daniel and me, our other staff people, such as the apprentice manager and product inventory manager, take groups for certain projects. We rely heavily on our staff to help watch for characteristics that will fit at Polyface.

Sheri and Teresa watch for etiquette and courtesies in and around the kitchen. They have their own litmus tests. Those of us who are outside with the candidates have individual pass/fail ideas. Obviously we need young people who aren't afraid to enter into conversation, but don't dominate. Balance, balance, balance. We want eager helpers but not space invaders. We want quick uptakes on instructions, but not so quick as to jump into the job before hearing the last of the instructions.

Any complaint, any whining of any kind, results in automatic disqualification. We purposely line up a pile of gut-wrenching, sweaty, hard work. Loading firewood, chopping brush, picking rocks, digging trenches, chipping branches, shoveling wood chips — these are all great projects. Who jumps out into leadership and sets the pace? Who holds back? Who hustles and who bungles?

If I died today, is this the person I'd like to replace me? I know this may sound morbid, but this is exactly what I'm thinking. In demeanor, spirit, and character, can this person run this farm without me? Not today, of course, but how's the fit? To ask for this caliber of individual is not a pipe dream. To you old curmudgeon farmers who I'm begging to open your hearts to young people, believe me, the good ones are out there. Yes, the ones that can replace you and run farther, faster, and longer. If you're going to invest in the internship program, you may as well invest in the best talent on the planet, don't you think?

Daniel's internal question is this: Do I want to spend the summer with this person? That's a great way to think about the process. It helps keep you from getting over-prejudicial about some mistake and helps you focus on the big picture. When you sit down to dinner at the end of the day, do you look forward to dining with this person?

Early on, we developed a rating sheet that looked like this:
1. Attitude — 30 percent
2. Aptitude— 20 percent
3. Awareness — 20 percent
4. Appearance — 10 percent
5. Acceptance — 10 percent
6. Articulate — 10 percent

If you notice, these add up to 100 percent. While someone reading may think that she can fool us with this inside track, forget it. These characteristics can't be gamed. Would-be interns either have it or they don't and the check-out exposes just about everything — including if someone is trying too hard to be someone he or she is not.

In recent years, we've moved away from this evaluation form to a more open discussion format. We've learned to trust our intuition, but perhaps that came about and seems to work well because we had this guide early on. I strongly suggest that you write down a description like this of exactly what you're looking for. Even if you don't use it formally, at least everyone will be on the same page.

During the two weeks, we do a mountain of laundry, prepare truckloads of food, have some terrific discussions, accomplish a pile of work, and meet some of the best young people on the earth. It's a huge Meet-&-Greet work party.

When we say goodbye to the last one, we huddle for the final selection process. Again, anyone of us has veto power. We've learned the hard way that if you don't have consensus, don't proceed. Each of us doesn't have to be equally gung-ho about every candidate, but everyone must come to an agreement. Like most of these situations, a few rise to the top as shoo-ins. A few disqualify themselves completely and obviously. It's the muddy middle that creates the interesting discussions. We haggle over it diplomatically until we come to consensus.

To facilitate this process, here's our procedure. Individually, we all make our top picks by answering the question: If I could hand pick my dream team, who would it be? Each of us comes to the table with that list. One person starts and we use a white board to write down that list. Then the second person reads his list, with duplicates from the first list receiving a tick mark on the white board. Third person, ditto. Fourth, ditto. Very quickly the top picks emerge with unanimous tick marks next to their names. Of course, others never get mentioned. This narrows the discussion to just a couple of names. Again, this session is fun and unifying; it's never been arduous.

We usually make our decision within a day or two of everyone leaving while all the faces are still fresh and memorable. A stale decision is a bad decision. As soon as we finish the selection, we send out the congratulatory acceptance notices — and the Dear Johns — to the prospective interns. Sometimes even at this late point, one will bow out. We've had interns who were so sure they would not be accepted that after they left the check-out, they went ahead and made other plans. Then when they got their acceptance notification, they were locked into something else. Bummer. It happens.

Sometimes we hold back on a couple of potential alternates who would normally get Dear Johns in order to get our commitment agreements before burning bridges. We give an immediate turn-around deadline of one week for the final commitments so if we're short, we can dip into the just-missed-it group and send another acceptance notice. It's never good to reject someone, and then later re-instate them — that's not professional. We do not keep a waiting list. Better to hold back on a couple of the rejects until you get final commitments from the candidates who made it. That helps avoid emotional yo-yoing for the would-be intern and underscores the professionalism of the program.

There you have it, the salient details for duplication. The main thing to remember is that it's okay to clearly and narrowly define what fits for you. The sooner everyone knows the page you're on, the better. And the less chaff you have to sort through.

Set your selection process in stone every step of the way. From application to your final selection each step needs to be dated and obeyed. This is ordered and professional. The process doesn't need to be cumbersome. If you follow these guidelines and rules, I'm convinced it will be one of the most enjoyable and anticipated farming activities of the year. Look at it as a person crop, rather than a pig crop or corn crop. Raising people has to be as cool as raising cane.

By way of inspiration, here's another except from British farmer icon George Henderson's 1944 *The Farming Ladder* on choosing apprentices. Notice the commonalities between his advice and what still works effectively today. He recommends starting at low pay and gradually increasing it, argues that higher education can be a liability, advises a check-out probationary period, requires attention to detail, and emphasizes that overall fit with the values and theme of the farm is critical to success.

How do we select them? An advertisement in the Farmer and Stockbreeder or Farmers' Weekly will bring in a good bunch of applications. Four or five who write the best letters are interviewed. No one is accepted without interview and suitable applicants are given the opportunity to come for a month on trial. Sometimes, of course, a boy is recommended by an old pupil or business acquaintance, but influence counts for nothing, he can only be accepted on merit.

What are the qualities we look for? Average height, build, and weight for his age; that he has not suffered from any serious illness or accident, for those who are taking up farming for the sake of their health have not the drive and energy necessary for the job; having worked on a farm during holidays, or kept small animals or poultry as a hobby, is a good recommendation; manual dexterity, indicated by a fondness for woodwork, or interest in science or biology — for agriculture is an applied science, also help. An enthusiastic Boy Scout, other things being equal, is almost sure of the job, for there is much in the scout training which is invaluable on the farm. We have always regarded Lord Baden-Powell as the greatest educationist in the world. Sometimes, of course, an applicant has had experience elsewhere for a short period, in which case we like it to have been a very good or a very bad place. If the former he can carry straight on, if the latter he will see the contrast. While a good general education is desirable, academic standards count for nothing. Of six boys, three of whom had obtained School Certificate, and three had failed, the last proved most successful. This is remarkable in view of the requirements for most trades and professions, and the examination is really only a fair test of general knowledge, though it does tend to select precocious children who can trot out the right little answer as required, against those who learn slowly but never forget what they have been taught. I think the real reason is psychological; the certificated thinks what a wonderful fellow he is, and is disappointed that the farmer does not share the same opinion, while the unsuccessful boy has been told so often by his schoolmasters that he is utterly useless, that he is gratified to find that the farmer looks for very different qualities which he usually has in full measure. In the same way the boy who has come in conflict with school authority is sometimes a great success, for he finds in farm work an outlet for his physical energy; while the good little boy who tells you he has never had a fight in his life usually gets homesick, and wants to run home to his mother after three days.

To have been to a farm institute or agricultural college is a disability which few boys can overcome on this farm; either they look on agriculture from the detached academic point of view, or they are looking for the opportunity for fooling about, which we will not tolerate. Our pupils usually feel the same. I remember overhearing one saying to another who was here on trial, "This job may be the only opportunity I shall ever have of becoming a farmer and I intend to make the best of it; if you want to play the fool go back to college where there are two hundred more like you."

Sons of farmers, business, or professional men have done best, and for that reason are favorably considered, though we have no social or religious prejudices. Individual merit is the only consideration. With sufficient money anyone can boast that he was educated at Eton and Balliol; only a boy of character and ability can say he learned his farming at Oathill.

We do not always find all the good qualities in any one individual; one does not when buying a horse. Horse-buying seems to me far easier, for in twenty years of farming I have only sold one horse for less money than I gave for it, while I have often been mistaken in selecting a pupil.

A pupil lives as a member of our family, with free board, lodging, washing, and insurance stamps. He receives ten shillings a week for the first six months . . .Not only do we pay good wages, but the learners actually earn them. A good boy after twelve month's training in our methods is far cheaper labor than any we can hire locally at half the money; without exception they all grow and put on weight, which would indicate that our high-speed methods have no deleterious effect on health.

Not only do we teach the practical work, but the scientific and economic aspects are carefully explained. It is quite common for a boy to acquire a better knowledge combining practice with theory in a couple of years, than by four years in an agricultural college. How much better he learns all the names of the weeds that grow in the fields, and the families to which they belong, if he is told them day by day while hoeing roots all summer; the points, bones, and organs of animals as he grooms them each morning; meteorology, if he is asked his weather forecast for the day at breakfast-time.

I always tell them the laborer should know how, the farmer should know why; that there is a reason for everything we do on the farm, and it can be given, whether it is the way to litter down a loose-box, or the order in which a horse's harness is put on; that a penny-worth of thought is worth a pound's-worth of manual labor; that the correct way is the easiest in the long run; that one should visualize the whole, but concentrate on the details; that everything I teach is the recognized standard practice of the best farms, and by following it they will be accepted as good farmers from Land's End to John o'Groats. For that has been my experience.

What a difference we find between these boys and the ordinary run of labor. How often, in the past, I have had a local boy doing something quite wrong and thought to myself, "Now I don't want to upset this chap, but I must go and carefully explain where he is wrong." Result? I get the reply, "If I'm not doing it well enough for you, give me the money due to me, and I'll clear out." How different with the better type of public or secondary schoolboy. He says, "I'm sorry. How foolish of me," and he does the job as he is shown.

To some people, of course, we would seem very fussy and particular in the way we have things done. But good work is only a matter of habit. We have all been told the old proverb, "Sow an act--reap a habit. Sow a habit — reap a character. Sow a character — reap a destiny.'" . . .

All the pupils we try are not a success. We can only teach when people are interested. We cannot help the lazy or dishonest, for there are some who would be quite happy to let others earn their bonus for them, and accept a free hand-out. But nothing is too good for those who are looking for the thorough training which will enable them to become capable and successful farmers, and who will help us run our farm as it should be run. We share our knowledge freely, will give any assistance they require to gain more experience or take a farm of their own.

While we prefer boys, we have had the pleasure of training others. A really capable man, who has made a success of some other career and is taking up farming later in life, can master in months the basic principles which it takes a boy years to learn, although he never achieves the manual dexterity of youth. Ex-officers and ex-servicemen generally, who are the most deserving of any assistance we can give, are terribly handicapped against business men. They have never had to think for themselves, everything is laid down in the King's Regulations or Admiralty Instructions, if any new work has to be taken up they receive a special course of training lasting perhaps several weeks, and they find it very difficult to adapt themselves to farm work and its entirely different outlook on life, where apparently one has no leisure, but has to study the theory and science while learning the practical work. The three fatal "S's" — Smoking, Swearing, Standing about — are as common in the services as they are on inefficient farms. To me they indicate lack of self-control, and betoken the man who cannot think or act without lighting a cigarette, the man who shows his irritation when things go wrong, and the man who cannot tell himself to get on with the job."

Chapter 10

Nuts and Bolts

T he internship at Polyface runs from June 1st to September 30th. No exceptions. We don't offer delays, deletions, or extensions. The reason is simple: we spend the first week in hardcore orientation, and we can't afford to do it twice. Everybody must show up the first day of practice.

We start at daybreak on June 1st. Most interns like to arrive at the farm a couple of days in advance in order to get settled in. Housing is dormitory style, with one building for men and one for women. We're not into throwing everyone into an old chicken house and letting them sort it out.

The Rules of the Game

Over the years, we've compiled a list of house rules. Different farms will have different house rules, but a program without any is doomed to failure because expectations are never defined. Silent expectations destroy relationships. One of our interns paid us the highest compliment during a media interview when she said she'd never been anyplace where the expectations were so clearly laid out and adhered to. That assessment tickled me to death.

Nuts and Bolts

Here are our house rules, pretty much in total, given to each intern upon arrival:

1. Sleep where assigned — no bed moving, hanging hammocks, pitching tents, etc.

2. Laptops and electronic devices are permitted for use in free time only. No phones, texting or ear buds allowed during work time. Wi-Fi is available from after your evening dismissal until daybreak chores. No Internet from daybreak chores until evening dismissal, excluding off work days. No use of Polyface phones.

3. Polyface provides safety equipment including earplugs and gloves. Lost items are deducted from your stipend. All interns are expected to have their gloves and earplugs on their person at all times. Keep up with your stuff. This is all part of your uniform.

4. Firearms are welcome. Discreet alcohol consumption is allowed. Be respectful of teetotalers and children. No profanity or smoking at any time.

5. We're a team. Guard against cliques and special buddies that can turn this into a soap opera. We've had four weddings germinate from our internship program. But romance has its appropriate time and expression. Get it?

6. Eating times. This has been an ongoing problem. Setting a mealtime limit seems to encourage using the maximum allotment every time, even when it disrespects a given day's project pressure. Some days are obviously more hectic than others.

If we don't set any time limit, some interns eat cheese and crackers and are back to work in ten minutes; others use meals as an opportunity to express their culinary prowess. So what is a mother to do? We've tried everything, from no time limit to X number of minutes per meal. Neither works well.

The lesson to learn is that the team is supposed to be aware of the big picture. You should be able to sense whether it's an unusually laid back day or an urgent day. We have more urgent days than laid back days (usually due to weather). Your advancement on the team, which includes being given more responsibility and frankly being enjoyed by the Polyface staff more, is directly a result of how you handle punctuality and demonstrate a git'er-done spirit.

Here are our recommended times: Thirty minutes to an hour for breakfast; Thirty to forty-five minutes for lunch, from entering your quarters until exit. Due to different work assignments, some interns will go to meals at different times than others, so it's up to the individual to watch the time. We know it's easy to turn a meal time into an impromptu world-problem-solving marathon; we enjoy this too. But that's for evenings. We make hay while the sun shines.

Meal prep and eating are confined to kitchens and dining areas, not yards, sales building and farm buildings.

7.   No overnight guests can sleep at Polyface or Polyface-leased farms. Visitors are welcome otherwise.

8.   Permitted activities: anything to make Polyface run more smoothly. All jobs are sacred.

9.   Chain of command: Joel, Teresa, Daniel, and Sheri are equal authority figures.   Second are the apprentice manager and full-time staff; third, apprentices. Interns are all equal unless otherwise designated.

10.   The sales building, packing room, and offices are work areas only. They're not for games, phone calls, entertainment, hanging out, parties or snacks. Use your housing for these activities. Don't leave any phones, gloves, hats, boots, or anything in the sales building area. This is our interface with the public, so keep it pristine.

11.   Evening entertainment (going to town at night) is fine, but don't expect Polyface to bail you out of a ditch when you've been barhopping at midnight, and you'd better show up bright-eyed and ready to work at daybreak. Polyface is not a babysitter.

12.   Drive s-l-o-w-l-y on our dirt road.   We push our relationship with our neighbors as it is with all the activity we generate and we

don't need you to be an additional reason for them to bad-mouth Polyface.

13. If you need to park at the sales building, park way away from customer parking during business hours so we leave the prime spots for customers.

14. The four-wheelers are not for recreation. They are extremely expensive to maintain and operate. Joel and Teresa farmed full-time for a decade without one; use your feet and legs. Feel free to run. If you can't do your job without the four-wheeler, then it's okay to use; otherwise, walk or run.

15. If you find yourself entertaining any negative feelings toward anyone or Polyface, it's your responsibility to find out the real story so you do not nurse ill perceptions. Rumors and gossip thrive in this kind of environment; don't be a part of it.

16. Personal appearance and grooming. Polyface is not the place to make a cultural shock statement with your personal grooming habits. Our customers need to be impressed with our cleanliness and classic appearance. The all-American look earns respect, especially when you're in the food business. Everybody, keep your hair clean and out of your face. Keep it like it was at your check-out when we selected you. This is not the time to let yourself go. Body odor will not be tolerated. Water is plentiful and soap is cheap.

All of these rules have grown out of problems over the years. We've found that it's better to be specific and address things in advance rather than leave things unsaid. You'd think some of it would not be necessary, but alas, alack, and forsooth, it is. Anyone starting up a program is welcome to copy these rules and put in your own farm name.

Interns are responsible for their own breakfast and lunch from the Polyface larder. They must prepare it themselves, but the provisions come from the farm stockpile. Their Monday-Friday supper is communal and prepared by our own on-farm chef primarily from our own on-farm production. The chef and horticulturalist work together to determine what is needed. The chef buys from the gardener at a pre-determined rate. That way we can track the cost of the meals and the gardener operates as a self-maintaining autonomous enterprise.

The day starts at daybreak, which fluctuates with the season, and ends at supper. Occasionally we have some pressing job after supper, but not normally. That is free time that interns can use as they wish. Two interns are on kitchen duty and they work out the rotation among themselves.

Two interns are on duty during the weekends, and they work out the rotation among themselves. Here is that fine line between top-down management and bottom-up responsibility. The farm staff, which includes the Salatins, apprentices, and full-time people like our chef, horticulturalist, apprentice manager, inventory manager — and others that may appear in the future that I can't even imagine right now — don't micro-manage the kitchen and weekend rotation.

We simply require that two people wash dishes each weeknight and two people show up for chores on Saturday morning. The interns have to work it out and police it themselves. And police it they do. Someone who ducks out, fails to show, or slacks in duties feels the wrath of the other interns. It's quite fun to watch them cover for each other and hold each other accountable. Usually one or two emerge as leaders who set the pace.

The interns trade around as necessary to do special things, whether it's attending a music festival in town or going home for a visit or touring a museum over the weekend. Sometimes they sleep all weekend. That's fine. All we require is that the two on duty occupy their stations. The interns appreciate this level of autonomy, realizing that ultimately it shows our trust in them. Nobody likes to be treated like a child unnecessarily.

Off time is when interns can catch up on emails, laundry, pet projects, reading or sleep. Of course, the full-time staff keeps things going during the weekends and after supper. One of the big differences between apprentices and interns is that the apprentices work our schedule, which is often daylight to dusk in the summer. The apprentices are always on call. Summer is sprint time; winter is rest time. That is one reason why we wanted the apprenticeship to be a whole year: experiencing the seasonal ebb and flow acquaints these young people with the farm's pulse. One of the big drawbacks to industrial-type farms is the unrelenting industrial work schedule.

If an apprentice came for only the summer, he'd think all we did was work. If he came only in the winter, he'd think farming was a piece of cake, a lazy man's way to make a living. Neither extreme is accurate; exposure to the full spectrum presents a balanced view toward the alternating fast-paced and then slow-paced rhythm of the farm, providing a more real-life experience.

A Lot to Learn

The first morning everyone shows up at broilers and we go through every step. We've found that the temptation to encourage the interns to jump in on the first day is a big mistake. It's much better to ease them in with lots of orientation. This nursery type situation sets up the rest of the season. We walk them through the whys and wherefores, the mechanics of everything, the ergonomics, and the efficiency.

Each intern receives a Standard Operating Procedure on the broilers (meat chickens). Lest anyone doubt how much there is to learn, I'm putting it in here in its entirety. I know this section has a lot of detail and some farm jargon, but I ask you to bear with me so you can experience for yourself just how much is involved in the simple command to "do the broilers":

1. Polyface field shelters are 10 feet x 12 feet x 2 feet constructed of pressure treated wood, one-inch poultry netting, and aluminum roofing. Uprights, both bottom ends, and open end left top brace are 2 inches x 2 inches, made by ripping 2 x 4s in half; braces are 1 x 6 ripped in thirds; all pieces 10 feet or 12 feet are 1 x 6s ripped in half.

2. Broilers per shelter may range from forty to eighty birds.

3. Grass should be no longer than eight inches; preferably two-to-four inches.

4. All shelters are moved and serviced daily before 8 a.m.

5. Mortality is kept at less than ten percent.

6. Birds must have feed and water twenty-four hours per day.

7. Bird weight at eight weeks should be greater than a four pound carcass.

8. Grit is placed in the feeder every Monday morning, enough to have some left Tuesday morning — normally about one pint.

9. All shelters are repaired within twenty-four hours of breakage; inspect for breakage regularly.

10. Prop shelters four to eight inches off the ground with a block of wood for additional ventilation when birds are more than six weeks old and temperatures are ninety degrees or more. Do not let birds get out of the shelter. (We now take off the center back panel in mid-June and replace in late August to eliminate the propping requirement.) The key is to be mindful of heat and prepare accordingly.

11. Be vigilant against predators — whatever it takes to control them: Guard dog, trap, shoot, sleep with birds.

12. A mortality rate greater than ten percent, except due to acts of God, is unacceptable.

13. Efficiency benchmarks:
    a. Moving: One minute per shelter
    b. Watering: Less than one minute per shelter
    c. Feeding: Less than one minute per shelter
    d. Field mortality: Less than ten percent
    e. Hurting birds during move: One bird per one hundred moves

14. When you see funky water in buckets and waterers, clean and refill them. Water must be clean. Funky water is white, with hair-like algae in the drinker. It occurs when the birds are very small, not drinking much water, and the weather is extremely hot. The birds

eat and carry bits of feed over to the waterer, which then ferment and grow a putrid-smelling, deadly bacterial soup. In the buckets, the culprit is normally green algae-covered insects. These reduce water quality and plug the hose.

15. Scrub all waterers and buckets between batches.

16. Place feeders specifically: parallel to the long side of the shelter and eighteen inches away from side. This facilitates birds eating on both sides of the feeder by giving them enough room to move around between the feeder and the side.

17. Waterers kept as near to the centerline as possible for two reasons:
>   a. Dolly prongs will not puncture the bottom of the waterer.
>   b. Weight will not make brace sag.

18. Water bucket egress always toward downhill side.

19. Waterers should be hung at beak height of birds until five weeks. Further upward adjustment is unnecesary.

20. Evening service from four to six p.m. Feed and water as needed. Take prop blocks (see number 10 above) out later if needed.

21. Only double feed when the feeders are completely empty in the morning. Do not feed more than twice a day. Feeding more frequently discourages foraging, thereby making the birds grow too fast and reducing taste, nutrition, and texture qualities.

22. Don't waste feed: use the scoop and never fill closer than one inch below the feeder lip. The birds will always dip it toward the edge. Fill feeders on the ground only, not on top of the shelter. Spilled feed on the shelter lid encourages crows and other birds. When the lid is lifted, excess falls into the waterer and soils the water.

23. Maintain feeder spindle at all times. Wood acceptable as replacement.

24. If dolly catches netting, repair at evening chores if hole not big enough for birds to exit. If hole big enough for birds to exit, repair immediately.

25. At least an eight-month rest and/or three grass shearings are required before re-use of the pasture.

26. No mortalities left in the field, ever.

27. Keep the doors squared and together on shelters.

28. Use plugs (varied size pieces of wood) as necessary to keep chicks in. If you see crows or buzzards around the shelters during the day, probably something is amiss. Often this means birds are out and being killed.

29. Shelters should be moved in a staggered formation without missing any ground; the trailing shelter should follow exactly adjacent to the preceding spot. Always keep a couple of feet between shelters to facilitate walking around. Move leading shelters extra if necessary in order to maintain enough distance. Always move the shelter a full length, so that the entire bottom is on fresh grass. Partial moves do not give the birds complete sanitary area or enough fresh forage.

30. Never mix feed and water buckets. Buckets are for dedicated use as either feed or water for the entire season. Store feed buckets upside down in the feed tank so they stay dry inside and don't blow anywhere. Store water buckets at the water source, and leave full so they don't blow around. We use colored buckets for feed and white for water.

Note: feed buckets are always colored and water buckets are white. Water dip bucket should be colored to differentiate it from carry buckets. Never dip water with carry buckets.

31. If feed gets wet, feed the wet material first before it gets moldy (normally two days).

32. Squirt enough Basic H in water to make suds when filling shelter buckets.

33. Carry two buckets whenever and wherever you go. Leave extra water by shelters and out of way of next move.

34. Take dolly to lead shelter when done moving.

35. Always use a sight stake to guide lead shelter so line will move straight down field. Move as needed. Do not cut off trailing shelters from fresh ground. If trailing shelters move onto soiled ground, move all birds and shelter elsewhere to insure new, unsoiled ground always.

36. Most common mistakes:

   a. Treating the shelters roughly. All movement should be noiseless and gentle. Don't drop shelters when pulling out the dolly. This procedure must be perfected until it is a gentle, seamless, smooth exercise.

   b. Running over birds. Move the shelter methodically and wait for the birds to walk away from the trailing edge. Short, fragmented motion is better than continuous moving. Have a child or dog work with you to encourage birds to move along if this is a problem.

   c. Failure to see breaks in the shelter structure.

   d. Failure to see ground depressions that let little birds out.

   e. Overfilling feeders.

   f. Letting water collect bugs and algae.

   g. Lollygagging.

Would you imagine that an instruction as simple as "do the broilers" could entail that much? This is not just a hustle-muscle deal. We want brains fully engaged at all times, eyes wide open at all times, ears straining for irregular sounds at all times. This is a full-on participatory farm — listening, observing, responding. Believe it or not, it takes four months to actually master all of these nuances.

For example, until an intern goes up to the field for evening chores and finds twenty little three-week chick carcasses dismembered, disemboweled and in disarray from crows and starlings, the admonition to plug the bottom edge in a field depression doesn't seem to mean anything. That little detail takes on new urgency once the intern picks up a bucket full of dead carcasses.

This is why at Polyface we have elected not to participate in short-term learning experiences. The level of awareness sophistication and skill necessary to replace the full cadre of industrial machinery, pharmaceuticals, buildings, and mortgages with people requires longer-term education.

In the first couple of days, we take all the interns on a farm tour to show them where everything is. Every farm has its ingrained jargon: the tool shop, the far flat field, the mountain pig pasture, pond A, etc. The interns need to get up to speed on this whole lexicon as quickly as possible. Otherwise you find them wandering around the fields without a clue where to go or what to do. Yes, we staffers sometimes chafe during these days, feeling like we're not getting anything accomplished. But the payoff makes the orientation worth it.

Chores are things that must be done at roughly the same time every single day. Anything else is not a chore, but just part of the workload. Non-chores can be fitted in throughout a day, sandwiched between chores. Like any farm, we have numerous chore stations:
- Move eggmobile
- Move broilers, feed and water
- Open feathernet nestboxes and feed
- Feed and check turkeys
- Feed and water rabbits
- Feed and water brooder
- Check cows

That's just for the morning. In the evening, another set exists:
- Move cows
- Feed, and water broilers
- Feed and water brooder
- Gather eggs
- Put away eggs

Some of these stations need two people, some need one, and some can use half a dozen. We've found that the best approach is to, again, let the interns choose their rotation on these, but stay with the same chore for a whole week. When they rotate within the day, it causes slippage.

For example, the broilers eat much less than a feeder full of feed and drink much less than a five-gallon bucket per day when they are small. In order to keep the water fresh and the feed from being stale or wasted, we don't want to give them more than they will eat in twenty-four hours. If they have lots of leftovers, it also adds to shelter weight, an ergonomic nightmare when moving them. We strive to have everything close to empty, but not completely empty.

If the broiler crew isn't consistent from day to day or day to evening, nobody knows how much to fill because nobody sees how much is left at the end of the day or the beginning of the next. To get a sense of the consumption flow, continuity must be maintained from day to day. We rotate chore assignments each week and that gives enough continuity to keep things right. More hands at the station doesn't necessarily get the job done better. It might be sloppier. It's also important for the interns to get a sense of the consumption progression and the work flow.

Although we certainly do want the interns to experience all the facets of the farm, rotating from facet to facet too quickly diminishes their ability to see the progression and the big picture. Keeping the interns on one chore for a week at a time enables them to get into a groove and grasp how things develop and grow.

Every Day's a New Adventure

Once chores are completed, everyone disperses for breakfast. After breakfast, the day's projects begin. The danger in writing this is that it sounds fairly cut-and-dried. In reality, each day is entirely different. One morning a week we have restaurant delivery load-up, so a couple of people come off chores to help assemble product. Three mornings a week we do buying club load up, where at least two interns and a staff member assemble the orders.

If a group of pigs needs to be moved, often we do that early in the morning during chores while the day is still cool. Pigs don't like to move in the heat of the day. Actually, the days unfold in a seemingly endless diversity of projects. We try to apprise interns of the work day flow so that as the week moves forward, we minimize surprises.

But as the old poem goes, "the best laid plans of mice and men" often require spontaneous responses to either a crisis or things that those of us in the inner circle knew about but failed to communicate to the interns. Just as in life, unforeseen things occur during a day that require attention. How we handle those things is a key part of the whole internship program.

Those of us in the leadership roles take turns with the interns, trying to create one-on-one times as much as possible. Perhaps one of the most misunderstood dynamics in a program as mature and sizable as ours is the assumption among interns that they will spend every waking minute alongside the farmer/owner. While that may be the case in extremely small, or fledgling enterprises, it is usually not the case.

"All you do is desk work and talk on the phone" is a common charge leveled by interns who don't have a clue how much administrative and managerial work it takes to shepherd an intern program. To operate a business of a size that can offer interns something meaningful for their time requires significant administrative oversight.

I don't know any farmers who enjoy this necessary sit-down work more than actually being out there sweating and making things happen. I certainly don't. But somebody has to order the egg cartons, placate the disgruntled customer, arrange for a special box of product for customer X. I've purposely not mentioned the government and insurance paperwork because I harp on that enough. Anyone running a business in modern Socialist America knows how monumental this is.

Every farm business that has progressed to the point of having a credible internship program started with long days in the trenches. As the farmer in one of these enterprises, I can assure you that the farmer has done his time. I've shoveled more dirt, picked up more firewood, dug more post holes, gutted more chickens, moved more shelters, toted more water and feed than any intern can imagine. The first year we composted our barn bedding after the winter — prior to using pigaerators — I hand shoveled the bedding into a pile that was forty-five feet long, eight feet wide, and five feet high. That's a pile of stuff to hand shovel.

As if that weren't enough, after it composted, I hand shoveled it into the manure spreader — some twenty loads. I would do about three loads every morning for a week. Today, we use front end loaders and shovel hardly anything. Teresa says we ought to require each intern to shovel one manure spreader load full just so these young people can appreciate it.

In the early days, I would do chores, eat breakfast, then head up the mountain in the truck to cut firewood. The truck in those days was a 1966 International Loadstar dump truck with a fifteen-foot grain bed on it. The sides were four feet high from the bed, which was nearly four feet off the ground. That made the sides nearly eight feet above the ground--a pretty high hurdle over which to toss pieces of wood. I would begin cutting by nine a.m. and usually have a load cut by noon. After a quick sandwich, taken while I sat on a stump, I'd pitch the wood up into the truck, much of it up over the eight-foot side, and rank the wood across the open back to hold everything in. I'd finish around three p.m., which gave me an hour to deliver it to a customer and get back home in time for chores.

In the early days, I'd get up at four in the morning and fill the scalder for chicken processing. It took an hour to heat up to one hundred and forty degrees. I'd go back to bed for half an hour, then awaken Teresa and the two of us would start processing at five. By seven we could do one hundred broilers. She would go in and get the children up, dressed, and fed while I did chores. At the end of chores, I would load up the next one hundred broilers.

I'd get back to the house by nine-thirty, usually, grab a bite of breakfast on the run, and we'd be back at processing by ten. The children were in play pens or helping as much as they could. By noon we'd have the next hundred birds processed, which left us an hour to clean up the processing shed and begin greeting customers at one in the afternoon. We'd handle customers all afternoon until five and then I'd go do chores while Teresa fixed supper. I don't need any intern telling me I've got a cushy job because I'm spending half the day administrating. We never hired a babysitter and worked even if we were sick. After all, customers were coming and we couldn't just call in sick.

In Sickness and in Health

This is probably as good a place as any to bring up a phenomenon that occurs every season. Easily eighty percent (that's a high percentage) of interns become sick for a day or two within the first three weeks. Only a couple have failed to get sick. We don't know why this happens, but our assumption is that most have never worked this hard. We don't take siestas. We don't take morning tea or afternoon tea. We go at it, full bore, all day, every day.

That's not to say we don't take a day off if the weather is bad or to celebrate the end of hay making. But generally, during the summer especially, we operate flat out.

During the first three weeks of the season, we watch the interns go down with sicknesses, one after another. Some go down in the first week; others take an extra week or two. But eventually most of them have their day or two in bed, usually with flu-like symptoms or sheer exhaustion. The ones who have just graduated from college and spent a lot of time sitting in front of computers are hit hardest.

By the third month, the interns harden off, just like a plant moved from the greenhouse to the outside world. Their physique moves to buff and their countenance glows with clear skin and joy. The transformation is palpable and never ceases to encourage me. To the person, they express the feeling that they've never felt healthier and more alive. What a joy to see these young people blossom. Even acne disappears. Sunshine, physical exertion, great food, and world-changing sacred mission offer healing on all fronts.

Prior to starting the internship, the only thing we offered was the year-long apprenticeship. Over the years, we calculated that the average male apprentice gained twenty pounds without changing waist size. In other words, that weight went to his shoulders, and upper body development.

Apprentices who come overweight shed extra pounds quickly. We've thought about billing the internship as an alternative fat farm. One apprentice who came a few pounds overweight was describing his transformation to a group of visitors saying, "I never knew my body would respond so magnificently." Yes, his sincere and serious line brought down the house with laughter. I guarantee you a summer as a Polyface intern will bring out the best physique you could ever want.

Women interns routinely complain of wrist and arm soreness for the first two months. The heavy lifting and toting strain and work tendons and muscles that have never been exercised to this extent. Just like any physically demanding program, it takes a while for the body to adjust. Just imagine if all that time at the gym could be channeled into meaningful farm work.

Extras

In addition to learning through working, the interns enjoy lectures and field trips that we sprinkle in throughout the summer. We do one evening lecture per month, and these have become highlights of the experience. I usually do talks on the basis and use of cow-days, designing and building water systems, and the gross margins on our various enterprises. Daniel usually does one on herding and sorting cattle and sometimes Sheri does one on marketing and social media.

When we do our Polyface Intensive Discovery Seminars (PIDS) the interns rotate through the educational experience just like the folks who have paid to attend. We offer three of these two-day seminars in July, literally shutting down everything except chores, to focus all our energy on being great hosts and teachers. About three interns per seminar get the two days off to enjoy all the benefits of the seminar. The others help prepare meals, do chores, ready wagons and all the other background things that go into hosting a seminar. Attendees also enjoy interacting with these interns. The interns often come away realizing how much they've learned. It's definitely a mutually beneficial arrangement.

Once every three years when we do the national Polyface Field Day, interns do prep work and act as co-hosts for the event. It's a huge event, with nearly 2,000 people attending from across the U.S. and even from foreign countries. We invite vendors we patronize to come (free) like a trade show. Barbecued chicken, beef, and pork along with fresh vegetables and chocolate buttermilk cake make the noon meal a gastronomical epiphany. All the interns get monogrammed shirts to help them stand out to the crowd. By incorporating the interns into both the fun times and work times they understand we're all a team. This creates appreciative loyalty within the ranks and is worth far more than the few dollars or hours invested.

We make sure the interns have a chance to ride with our delivery driver on buying club drops and restaurant deliveries. We usually spend a day at the slaughterhouse we use and co-own, T&E Meats in Harrisonburg.

We rotate participation in food fairs and sustainable agriculture events. On their own, of course, the interns arrange their own excursions to see farms, museums, or other places that attract their interest. Since many come from west of the Mississippi, they routinely organize a weekend to visit Washington D.C. Of course, two must stay behind to help at the farm, but the group works it out. Usually at least two are from the East Coast and have already done the Washington D.C. thing.

The idea is that within the farm work context, a good internship program includes a healthy amount of formal teaching and off-farm learning venues that round out the experience. We emphasize, though, that the interns earn these plums as a result of being faithful in their work. You can't fool around with the labor component and expect full ancillary benefits. As mentors, we must deliver these gracious components, happily and magnanimously. This creates an environment of mutual loyalty and appreciation.

Chapter 11

Housing

Perhaps one of the biggest conundrums of any internship program is housing. I've touched a little on this already, but I want to drill down into this issue now because in contemporary American culture most people don't live where they work, and most employees don't provide on-site housing to their employees. In our culture of wide open frontiers, personal space counts, making this a critical issue of concern for mentors and interns alike.

As much as you might think it would be fun to have interns live with you, it simply doesn't work. Everyone needs some space. Farm couples need to be able to argue. Interns need a place to retreat to as well. I'm the most affable people-lover in the world, but I still want to have a place where others need to knock before entering — my private space, my sanctuary. It doesn't have to be big or spacious. A hovel is fine. But I need it.

A temporary live-in situation is tolerable because everyone knows it's temporary. Obviously if you have a self-contained, separate-entrance apartment in your house, that's probably okay as well. A big farm house remodeled to independent living areas can work. I'm not saying the housing has to be under a separate roof. But each living space needs to be lockable, private, personal. You can't simultaneously have unfettered access throughout the quarters and still preserve the sanctity of individual space.

So how do you make that work?

One of the biggest enemies of farm internship programs are building inspections and zoning requirements. I've already addressed this, so let me go on with some alternatives.

Our county — Augusta County, Virginia — has five exemptions to its building codes. These include structures that *can* be placed on agricultural land without getting a special use permit. In other words, they are permitted structures rather than prohibited structures. Prohibited structures *can* still be built if you can convince the appeals board to grant a special use permit. Just because it's prohibited doesn't mean you can't do it; you just have to get permission.

But the beauty of the five permitted structures is that you don't have to get permission in order to build them. While these same approved building types may not exist in every jurisdiction, I'll bet some permutation does. You should at least be willing to explore your local options.

The first option is an agricultural accessory building. Typically these are considered to be barns, equipment sheds, or farm shops. However, if my farm requires people, why is a structure that houses drying garlic or pooping pigs more an accessory to the farm's production than the people to run it? I find it fundamentally absurd, if not downright hateful, that I can amass as many tractors, machines, vegetation, and motors as I want without any permits at all. Oddly, I can replace all the people with machines, and that's completely blessed by the local zoning board.

But if I decide to replace machines with people, suddenly that's anathema to farming, at least for contemporary zoning officials. I can even build houses for thousands of chickens, thousands of pigs, thousands of cows, without any permission whatsoever. They can be as large as I want. But put a person in a tiny shack, and suddenly it's a licensable use of the land.

The biggest political problem here is the open space preservationist who typically comes in with the heavy hand of regulation, assuming more often than not that people are bad for the land. In order to curb suburban development, their open space easements and preservation requirements impede a human-centric food production model. They push for a legal paradigm (a zoning structure) that suggests it's much better land management to populate the countryside with machines than with people. Unwittingly, these policies create barriers to bringing loving stewards back onto farms. Those of us replacing machines with people find that, ironically, our biggest practical implementation hurdle is the constituency that wants local ecological food while pushing for policies which impede its production. Talk about intellectual schizophrenia.

Oddly, these preservationists actually end up aiding and abetting Big Ag--corporate, industrial farmers. The unintended consequences of their actions could not be more hurtful to the very kinds of small and ecologically-based farmers they ostensibly want to encourage.

Honestly friends, it's time for some sharp attorneys to take this issue to the Supreme Court. They should argue with gusto that housing for tractors is far less necessary to farming (on multiple fronts — jobs, ecology, education, local economy) than housing for the people who drive the tractors, drive the education, preserve the ecology, and serve the local economy.

On the face of it, it's absurd that a tent for interns is considered by zoning to be more invasive to green space than a concentrated animal feeding operation (CAFO), or the shed to house a two-hundred horsepower articulated eight-wheel drive tractor! Most states have some permutation of the American Farm Bureau-supported Right-To-Farm law. This was developed to protect farmers in the late 1970s from nuisance suits like noise, odors, and clogging rural roads with big equipment. Amazingly, nobody thought it was necessary to protect the farmer's right to house labor on the farm.

In my travels, I've seen numerous barn lofts and outbuildings turned into intern housing. I have a friend who says every farmer should embark on a massive farm building program to get the roof on the map. At some future date, you can turn the building into a house or bed and breakfast, restaurant, hospitality suite, or intern housing. Once the roof is established as part of the landscape, nobody really cares what you do underneath it. I tend to subscribe to this view. Generally, the people who retrofit an existing farm outbuilding to intern housing pretty much fly under the radar.

Another silly nuance of this is that you can have unlimited guests in your house without question. A farm family could have ten interns move in with them, in their house, for an indeterminate period of time, and that would be perfectly legal. But if the family decides to build a tiny cottage under a separate roof for just one intern, that often violates numerous codes. This is all absurd, and of course is one of the biggest reasons farm internship programs are not more abundant.

The environmentally-minded urbanites who want local integrity food to proliferate need to elect some libertarians who will roll back these ridiculous laws so that farm mentorships and internships can grow. Too many of these foodie-focused urbanites bask in the self-righteous assumption that they've preserved open spaces to produce local food when in reality they've erected massive hurdles against their farmer friends.

Farmers, go ahead and erect as many sheds and barn-type structures as you can. Call it an investment in your future. You can always remodel, retrofit, and re-commission them as the need arises.

The second building exemption in our county is for tree houses. I haven't been able to ascertain for sure if all four corners of the house need to be in the tree or if one is enough. But this one certainly intrigues me as doable wherever you have a cluster of trees. The law doesn't say how high it needs to be. Neither does it say it all has to be in one tree.

If you had a group of trees then you could conceivably affix floor timbers to the trunks and build up from there. This obviously has the problem of injuring the trees. What if one dies? Again, I would argue that what is important is that you build it today within the parameters of the rules. Nobody, five years down the road, is going to tell you to tear it down if one of the trees dies and you have to put a concrete pillar under that corner. Very few of these rules apply to structures already established on the landscape. As long as a bureaucrat signs off on it today, or as long as it passes the code today, that's all that really matters.

Building inspectors never come back to houses they passed and do a re-up in ten years. It's all about pass-fail today to today's standards with today's building police. Have you ever seen a building inspector racing around a decade-old subdivision re-inspecting houses? Of course not. This is one of the little-discussed nuances of the building licensure system.

I have a friend in Australia who is quite a knowledgeable advocate of composting toilets. The problem is that the small models certified by inspectors don't work very well. We've wrestled with this in our own intern housing and finally abandoned them. You should either use a bucket and cart your waste out to a functional working outdoor compost pile or build a composting toilet with a monstrous carbon chamber. He suggests that you go ahead and buy the approved non-working model, get your occupancy permit, then sell the toilet on e-Bay and use a 5-gallon bucket with outdoor pile.

In the case of tree houses, the fact that your house might eventually kill the tree doesn't negate the fact that today it's a tree house and therefore exempt from building codes. Call it a playhouse for the kids. Call it a super hubba-bubba dog house. With your best sad-eye expression, tell the regulators your wife demanded that you build a man cave for all your masculine hobbies. She threatened to divorce you if you didn't build it, but promised to love you and raise your children and keep the kids from using drugs if you built your tree house.

If you think I'm advocating lying to game the system, yes, I am. It's the kind of lying the Hebrew midwives did to Pharaoh when he ordered them to kill babies. It's the kind of response Jesus had to the Pharisees when he drove them from the temple with a whip for desecrating His house. As far as I'm concerned, land stewardship, or if you prefer, creation caretaking, is as much a mandate as saving babies and reverencing the temple. If some bureaucratic goon or misplaced public policy outlaws what is righteous, then I adhere to a higher law.

What's neat, though, as far as I can tell, is that a tree house is totally legitimate without any explanations whatsoever. You don't have to explain its intended use.

Interestingly, the leverage factor the government uses is the electric hook up. If the structure can be built without a new electrical drop, your options increase greatly. For a few thousand dollars you can bury an extremely heavy electric line a long distance from an existing source. Even if a thousand feet cost five grand, that's a small price to pay to keep from visiting a bureaucrat. It's some money, but far less hassle in the long run than trying to punch through a phalanx of building police. And what if it's totally off-grid? Maybe your area has good wind or solar power? Maybe a stream you have can provide hydro-electric? Maybe you can completely fly under the radar? If you can, why not?

The third exemption is if it's on a chassis. Although this is written primarily for recreational vehicles (RVs), it applies to anything that's portable. Some codes use the term "permanent foundation" to help define what fits under this rule.

This provision was what enabled us to build our first apprentice cottage, which we put on two I-beams affixed to four concrete pillars. That way we could honestly say it was portable. We called it a farm machine. If the only thing that distinguishes a regulated structure from an unregulated one is portability, then make it portable.

In a day we could slip a hay wagon under it, jack the I-beams off the concrete pillars, and move that whole cottage to another site. The fact that it wasn't as easy to move as a hay baler did not fundamentally change its designation as a farm machine. We used a composting toilet and ran the grey water out into a little pond of cattails and irises. It has worked great for twenty years.

Over time, the apprentices added a mud room on one side. Later, we added a sunroom on the south side. Both of those are on the ground, but again, remodeling with porches and sunrooms on an existing structure is not very noticeable. The point is that we established a structure — albeit semi-portable — on that site. Once people get used to seeing it, modifications are academic. For the most part, nobody notices and nobody cares.

I would much rather have a few well-kept, semi-portable cottages dotting my farm than the typical multi-machine bone pile of rusting machinery that is completely legal and common. Which is really the bigger eyesore? That farmers like me have to spend significant time ferreting out these grey areas just so we can teach consenting adults how to farm is both outrageous and evil. On top of that, we live daily looking over our shoulders lest some bureaucratic zealot swoop down with nothing better to do than harass a vibrant, profitable, working farm over different definitions of how to run said farm — my farm — which I'm running just fine on my own, thank you very much Uncle Sam!

One season we bought a small camping trailer and one of the apprentices lived in that. We parked it down by the woods and it worked beautifully. Granted, it's not something a person would want to live in for a lifetime, but for a few months or even a year of training, it's fine.

On this note, yurts are not considered buildings in most parts of the country. For many intern programs, yurts are a perfect non-building. Yurts, as far as I can tell, are never classified as buildings and are therefore one of the quickest and simplest means of providing intern housing. The problem with them, of course, is that internal walls are difficult. They don't have structural timbers to attach anything and they are round — both possible complications when you're putting in walls for bathrooms, kitchens, and bedrooms.

We actually looked at yurts but realized that since we had plenty of forest for lumber and our own band saw mill, it was silly to spend that much on a structure when we could build something more substantial with a lot less money. For many farmers, however, yurts are a perfect answer to the intern housing problem since they are considered a non-building. Kits are readily available and more builders have experience building the pad and erecting the yurt. They are simple to heat and even come with gutters. With today's plastics technology, extremely transparent windows can be sewn into the covering for additional solar gain. Yurts are interesting structures with lots of space inside requiring relatively little materials. The fact that they aren't buildings simply adds to their charm and desirability.

When we began our internship program, we offered it to six males so we wouldn't have to deal with two structures at once. We could at least start the program, and then add females later. A local church had used a trailer with three Sunday School rooms in it while constructing a new educational wing. With the addition completed, the church didn't need the trailer any more. Nobody would take it so it was moved onto a neighbor parishioner's property.

In a couple of years the tires dry rotted and went flat. Bushes grew up around it. Then the parishioner wanted to sell the property and needed it moved. By that time it was fairly unattractive — just the kind of thing you'd want to house interns. That's a joke, dear people. The owner wanted to give it away to anyone willing to remove it from the property. We looked at it, could find no leaks, and the price was right.

We jacked it up, removed the wheels and replaced the tires. We hooked it up to our pickup truck — it was light but bulky — and towed it home. That first group of interns spent their first week installing the kitchen and bathroom (remember, it had just been used for Sunday School rooms). We got three pairs of bunk beds and some mattresses and for very little money, we had a bunkhouse. We called it the Roost. A composting toilet, power, and water completed the functional facility. It was on a chassis, so it was legal. Just a farm machine. Some machines plow the ground, some bale hay, and some house interns. We could house them on a plow or baler, but this seems to work better. You get the drift.

As long as you have a non-building, you don't come under the codes. The trick is to find something that's a non-building, whether that be a tree house, portable, or whatever.

The fourth exemption: if it floats. Seriously. Our county has no large bodies of water and so no ordinances regarding floating buildings or houseboats have ever been written. It's simply a non-regulated area. Several years ago I came very close — my family would say too close — to buying a houseboat and floating it in one of our ponds for the female intern housing. I thought it was the coolest idea going until I learned that houseboat dimensions are based on the floating footprint.

One of the distinctive characteristics of houseboat design is that it maximizes deck space and minimizes living space. Why does a person live on a houseboat? It's not to stay indoors, but to sit outside on deck chairs sipping lemonade, communing with sea gulls, and soaking up sun rays. The bottom line is that a 10 x 24 houseboat has about has much living space as a 6 x 12 camper. That's too tight.

As we were sleuthing this option, you can only imagine all the jokes about the women being out in the middle of the pond. The guys would have to cross a moat to get to them. Each of our ponds could have a houseboat on it. I thought it was a hoot — everyone else in the Salatin family thought it was a horrible clutter. Contrary to popular belief, all of my ideas are not considered gifts from heaven even by my own family. What are families for? They're to keep us human, I guess. While my readers may sometimes think I'm pretty clever, my family knows better. Ha!

Don't think I've given up on the idea. Just the other day we purchased some 2,500 gallon food grade plastic cisterns. Can you imagine putting a wooden frame around two of those? Let's see, 5,000 gallons would displace 40,000 pounds of water. That sounds like a nice little cottage to me. Who knows? We may be forced to custom-build a floating intern house yet. I think I have a new idea. Forget the houseboat. Float it on some tanks. Goodness, you could even use a bunch of barrels.

Imagine a raft of a hundred plastic barrels held together by a wooden frame. The floor joists would go right across the barrel tops and you build up from there. At first, it would float way out of the water, but as you built the structure, the weight would gradually push the barrels down. In just five feet of water, you could float a 40,000 pound house. Even a very large house trailer doesn't weigh that much. Oh my, the wheels are spinning.

Did I mention that every time we get a few thousand dollars, we build a pond? You can build a pond big enough to float the structure I've just described for less than a few thousand dollars or so. Goodness, you can easily have that much tied up with foundations on a regular building. For supper, you just dangle some hooks out the windows and eat whatever you catch. Put the bathroom upstairs and run the sewage out a pipe to the bank. Better yet, use a composting toilet bucket and have a floating compost bin adjacent to the house. The flies come to the humanure and lay eggs, which hatch into larvae, which fall into the water, which the fish eat, which you eat, which makes more humanure . . . . This is an unregulated structure, so you can do whatever you want.

I wouldn't flush into the pond, although if it had enough carp and catfish, it would probably be fine. That's what they do in Vietnam — toilets perched over the river. If you're not quick at getting up, the catfish whiskers tickle your bottom while the lips... Probably enough of that for now. The point here is to not let the tyrants stop us. Let freedom ring.

Finally, the fifth exemption: a hunting camp. Yes, a hunting camp, as long as it's no more than nine-hundred square feet. By the time this book goes to press, our farm should be fully utilizing the

Polyface Hunting Camp. Interestingly, the law doesn't say what you have to hunt, so at our camp, we're hunting for the truth.

Built with Polyface lumber, our truth-hunting camp incorporates a sleeping loft and two awnings (wings): one houses the apprentices and one houses the apprentice manager. A nearby camp for the female interns allows us to concentrate the infrastructure around power and water. The ticklish part of this project was getting power without a building license.

Prior to any construction starting, we built a pond — between the two camps. After all, we don't want it to be too easy for the guys and girls to get together. You've got to have at least a little adventure in the hunt, right? The electric company routinely installs power drops for livestock watering systems. We registered this drop as a livestock watering system — which is totally honest. We've had that pond site picked for a long time. It's in an old gully, but high up on a hill that will create considerable pressure leverage for pumping. We can install a pump on the edge of the pond and the hunting camp is just accessory use.

The power drop was free, of course. All we have to do is pay the monthly bill and we're all set. It's a pretty nice two-hundred amp installation for a water pump, I must say. You just don't know how big a pump you may need down the road. A recycled grey water system cuts water use in half. We've installed six 2,500-gallon cisterns, for a total of 15,000 gallons accumulated from roof run-off. The idea is to be completely water self-sufficient. Eventually we want to erect a hoop house for the grey water, running it through carbonaceous troughs that grow hydrologic plants. This is a permaculture idea that has been done throughout the world. If we can make it work, we'd love to use an algae treatment on the sewage. That way the whole system would be entirely self-contained, without pulling water from off-site nor discharging sewage anywhere.

By the time you read this, I may be in jail because of this structure, but I hope not, since it's *only* a hunting camp. It's tucked in behind a hill so it's not visible from anywhere. We've purposely used a barn-type design and earth-tone metal roof and siding to blend into the landscape. Our daughter, Rachel, configured the inside to maximize functionality per square foot. We purposely made the sleeping loft too low to be counted as habitable space. That way it doesn't count in the square footage.

The housing, throughout the years, has been a major cost and investment in the internship program. After we installed the Roost for the guys, I was totally crucified everywhere I went to speak for not accepting girls. We don't believe in cohabitation, so in order to have both sexes, we needed two housing options. I came back after that first year and begged the family to figure out how to take females.

On the speaking circuit I was accused of being a dirty old man, a sexist, a woman hater — you name it, I was just guilty-as-charged in the court of uninformed public opinion. At the time, we received three or four female applications for every male. We decided we had to take females. But where to house them? We found an apartment about five miles away. It was the upstairs of a two-unit complex and looked roomy and perfect. But at $4,000 for the summer it wasn't cheap. We decided to try it and opened up our application process to females. Only one applied. Then a second one. We took them both. No other female applied.

In fact, after opening up the program to females in 2008, we took every single one that applied for the first three years. Now, how do you explain that it went from four females to every one male applicant to next to no female applicants? It seemed like sour grapes. Women were previously just applying to make a self-righteous point, without bothering to explore the real reasons behind our process. Fortunately we weren't (and aren't today) a public institution receiving government money so we could do what we wanted and not be sued for discrimination.

The first two women interns were spectacular and we decided we wanted to continue. The next year we found another place nearby but the women were concerned about not being on site. "We have to get up ten minutes earlier than the guys to get here and start chores, " they complained. They also were miffed at having to buy gas to get to the farm, even though that was made clear as part of their application process. That year too, we had some issues with the guys, so we decided to move the whole bunch off premises. The entire thing has been a process of experimentation, evolving through trial-and-error. But that keeps us on our toes, right?

The next year, 2010, we rented an upstairs-downstairs duplex about five miles from the farm. Girls upstairs, guys downstairs. That worked, but it was $10,000 in rent and unhandy. It did give us our peace and quiet, though. Although it was an extremely spacious living arrangement, the interns yearned to be on the farm full-time. They wanted to live here.

The next year we built a separate cottage for the women and housed the guys in the Roost (the Sunday School trailer). That worked best and we could see the results in an improved team spirit. Things were coming together through immersion. By that time we added the communal evening meal, too. The combination of one daily communal meal and housing on-site created the magic for the best season so far.

I highly recommend on-farm housing, although off-farm sites will work in a pinch. It takes some extra effort to create social times to compensate for off-site housing. And there's no replacement for being here when those unexpected farm crises occur. Those moments offer the most audacious learning opportunities through sheer immediacy and the call to action.

This current arrangement — essentially two bunk areas — creates an interesting nuance for couples who want to come. Of course, sometimes families want to come and intern. But because the team spirit is so vital to the whole experience, as of this writing, we don't accept couples for the program unless they're willing to live as singles. We can't have a couple in either bunk facility. We don't want them in our house. There's no place else to go on the farm. If they live off farm, it creates a significant separation from the rest of the interns and creates a "them" and "us" vibe. And, of course, couples would want compensation for off-site housing. After all, they reason, "you're housing those guys. How about us?" But that just isn't practical or affordable in an already tight internship program budget.

We've already invested in the housing infrastructure for the number of interns we want. Why should we drop those on-site interns and pay cash for off-site housing? That's a direct added expense. We're glad to take couples, but they must agree to live as singles and most don't. The only couples we've worked with are when only one of the pair comes. This is no longer than a military deployment, and people do it all the time.

While this may sound unnecessarily harsh on our part, we've tried all the combinations and discovered what fits for us, for *our* program. And here's the bigger point: To mentors, I say feel free to find whatever fits with *your* life, *your* farm, *your* family, *your* resources — and don't apologize. Lay it out the way it has to be in order for you to maintain your sanity and your finances. Whoever doesn't like it won't be an intern, and that's fine.

To you interns who think ill of me for creating our own template, just because we have one template doesn't mean you can't find one that fits you elsewhere. What if I only wanted Pakistani atheist bow-legged vegetarians as interns? If that's who I like and it fits, who cares? Maybe you'll find someone who only wants people just like you. Or maybe you'll find someone who doesn't care about anything. I know some farmers just throw all their interns, guys and gals, into an old chicken house and say "have at it." It's their farm; they can do as they please in my opinion. Plenty of choices exist.

We appreciate diversity, but too much in your face every day can be a drag. As an intern, you don't want to be a drag, so go where you fit and don't demonize or speak prejudicially about a place where it's hard to fit in. As mentors, it's our place, our house, our farm. That's true. But if you can't keep interns or if some leave before the internship is over, chances are you're either too picky or you're too difficult to fit with anybody. It takes two to tango.

House Rules

You'd think that young people at least eighteen years of age would know how to live in a house. Wrong. Don't assume anything. You have to spell it out. Here's Teresa's summer housing protocol for the interns:

1. White mattress pad stays on the mattress all summer.

2. If using our bedding, the fitted sheet, flat sheet, and pillowcase must be used.

3. If using our bedding, it must be washed at least once a month.

4. If using your own bedding, please keep a fitted sheet on the whole summer — if you can stand sleeping on the same bedding all summer without washing it, fine by me. Please return our bedding to the plastic tubs in the hallway.

5. The bedroom area must be cleaned (vacuumed and clothes picked up) at least once a week.

6. We have trouble with mice and ants so . . . keep all food (especially the sweet and sticky) in an airtight container or in the refrigerator.

7. Rinse all dishes immediately after use if you can't wash them immediately.

8. All dishes should be washed within twenty-four hours of use.

9. Get rid of garbage (food and table scraps) each day. You can have a bucket outside the door to put scraps in and then dump the bucket frequently. Those scraps can go to the chickens.

10. Likewise, the burnables and recyclables need to be dealt with in a timely manner. At the end of the summer, all burnables and recyclables need to be removed from your housing areas and taken to the proper receptacles.

11. Keep the counters and tabletops wiped clean after each meal. (Mice and ants, again.)

12. Vacuum all floors at least once a week.

13. Mop the bathroom and hallway floors when needed (or maybe more often).

14. The shower and the sink in the bathroom will need attention often because of the heavy use. Every week they need to be cleaned thoroughly.

15. Leave muddy boots/shoes by the door; cleanup will be easier.

16. If anything is broken or not working properly, please tell one of the Salatins as soon as possible. Keep reminding until we deal with the issue.

17. Polyface will supply toilet paper, dish soap, and cleaning supplies. Let someone know when you need more so we have time to purchase those items before you run out.

18. You are living with other people in a small space — be respectful of them and their space. The more quickly dirt and clutter are dealt with, the smoother the summer will go for everyone.

19. There will be random inspections of your housing areas, so keep them clean.

20. Clean out the refrigerator monthly and throw out rotten food.

21. The Polyface walk-in cooler is off limits for intern food storage.

I'm sure this list evoked some smiles among mature adults. But the immature adults likely don't have a clue. Do you know what a mattress slept on by an intern for the summer, without any sheets above or below, looks like at the end of a season? Do you know what a pigsty looks like? Smells like? Like other protocols listed in this book, every single one of these has a story behind it. Did I say we've been at this for a long time?

Lest you think all young people are slobs, we've had some real clean-freaks over the years. One intern practically white gloved the Roost every day. He grew up in a home where if the vacuum streaks weren't clearly visible on the carpet, it was time to vacuum again. Another tidy one came to us directly out of the Navy, where he was a captain. He sure kept things in shipshape. But these are the exceptions, not the rule. You can't build your protocols assuming that young interns, some of whom are on their own for the first time, will know how to live and keep a respectable house.

This is all part of the learning experience, part of the presentation. Everything relates to everything. I don't want to buy food from a dirty person. But I can assure you that I'm more than a bit concerned about the cleanliness of the food when the farm family lives like slobs. An ordered home indicates an ordered farm. An ordered farm implies clean food. These things go hand-in-hand and it's important for the interns to understand these more esoteric nuances of success. And like other aspects of what a mentor models — frugality, profitability, the willingness to jump in and perform any job — requiring cleanliness sets the interns up to expect cleanliness and orderliness in their future farm pursuits.

All in all, I recommend housing on or as close to the farm as possible. If it's your building, clearly defined protocols are a must or it will be trashed beyond recognition within a year or two. Visit the housing often--pop in when everyone is out in the field and look around. We've found interns pretty forgiving about amenities but have generally found them unappreciative about how primitive some may have it. As we've upgraded, we haven't found a corresponding sense of appreciation. The point is I can't imagine any reason to create a Taj-Mahal housing space for interns. Make it functional and respectful, but keep it rustic. No matter what you do, housing will be your major intern expense.

Chapter 12

Leveraging the Labor

Believe it or not, one of the biggest weaknesses in farm internships is actually keeping interns busy. Several things account for this and the solution is not one-sided. At Polyface, we've wrestled with this issue and believe both the problem and solution are a shared responsibility between mentor and intern.

First, most farmers aren't used to working with others. Farmers don't trust interns to carry out the jobs correctly. As a result, mentors tend to withhold a lot of tasks and not delegate.

In these cases, while we can certainly blame the farmer for keeping the work too close to his chest, the intern needs to step up and volunteer. An observant, with-it kind of intern shouldn't be clueless about what needs to be done. In the intimate relationship that inevitably develops between the mentor and intern, I can't imagine a case where volunteering is inappropriate. While it may not be appropriate for an intern to volunteer for skilled tasks, at least at the beginning, something is always within skill level, even if it's simply to hold the shovel.

While I agree the bulk of the responsibility for keeping interns busy is up to the mentor, some mentors may need to be helped along with this. The reason I bring it up this way is because in our own experience this has been a bit of a problem. When you go from a workforce of five people one day to twelve the next, for example, that's a lot to assimilate. It's especially hard when more than half the crew can't be trusted yet to do much of anything.

While I like to see people busy, I've had to mellow to the point where I'm not frustrated when interns aren't working — in some situations. Many jobs don't require two people. Farmers, accustomed to working alone, often enjoy continuing to work alone even when interns come along. The comradeship of a second person is often offset by the extra questions and space invasion it engenders.

I'm sharing this burden because I think it's important for interns to understand how it grates against a farmer's psyche to take unnecessary people to a job site. But taking extra people is how the information is transferred. I can certainly string a cross fence for the cows by myself — I've done it all my life. But I've learned, slowly and reluctantly, that I need to take an intern routinely in order to teach him how to do it. If I don't bring him again and again, I'll come to the end of the season with a lot of intern cost and no payback because the interns will have never developed the skills to do jobs on their own.

When necessary, I'll even say, "I don't need your help on this, but I want you here to observe and see. Don't feel like you have to help." Verbalizing that helps take the pressure off both parties.

I'm a huge advocate of indicating, by body language and positioning, that you want to help. "Keep your hands out of your pockets," is my refrain, as my children will tell you. One day we were rebuilding the awning on the far side of the barn when Daniel was about ten years old. I had a friend helping me and we weren't getting along well with the project. I can't remember what the problem was, but we had trouble getting the rafters aligned and connected and I was getting frustrated.

Teresa came out from the house to check on our progress, and Daniel met her at the corner of the barn because I was on the tractor. When Teresa inquired of Daniel how things were going, he announced forthrightly and out of my hearing, "Whatever you do, Mom, just keep your hands out of your pockets and everything will be okay."

I confess that it's difficult for me to tell an intern, "Don't help, just watch." Something about that vexes my soul. But it has to be done. The proper intern response is not to disengage, but to be as involved as possible without interfering. Don't go chasing after butterflies and asking questions about the forestry program in China. Stay with me, right there, within an arm's length, so that if I collapse of a stroke, you can pick up the spool of wire before it falls out of my hands and breaks on the ground. Act interested. Be there. Engage.

As an intern, you need to be willing but not pushy. Often it's psychologically awkward for a farmer to have a second person along. So help him out and act interested and eager. And don't be blind about little things that can be done.

I'm known around Polyface now as the junk man. That's because I seem to be the only one who can see the cardboard box along the driveway that a gust of wind picked up from somewhere and dropped as litter at the farm entrance. Sometimes I purposely walk past it for a day or two to see if anyone else will pick it up. Usually not.

Why am I the only one that sees these things? The little clump of thistles behind the equipment shed — am I the only one that sees them? When the interns are all out lounging on the lawn waiting for instructions, where is the one who will see the box or thistles and take care of it? An intern who can't find something to do will probably not be a successful farmer.

The second reason leveraging intern labor can be problematic is that we farmers aren't used to sharing everything that's in our heads. When I get up in the morning, I check the temperature and look out at the weather. Much of what we do is weather dependent, which means it can change from one hour to the next. A good farmer has a list of contingencies in mind.

If it's hot, we do this. If it's cold, we do that. If it's raining, we work here. If it's sunny, we work there. If it's windy, we can't do this job. If it's still, it would be a good day to jump on that project. If it's still and sunny, here's what we need to do. A veritable cascade of these things are dancing around in farmers' heads and we're simply not used to saying much about it. We're just ready to do it. This is just part of orchestrating the day and making things happen.

To entrust all of that material to the interns is tedious. We have to force ourselves to share the bare minimum so people know what's going on, realizing it could change in the next hour. So how much do you tell? If you tell it all, it's like farming by committee and discourse. Nothing will ever get done. At Polyface, we're always trying to strike that balance between the need to know and just "follow me." It's not easy to strike that balance and probably nobody does it perfectly.

Interns generally want to know more. Farmers tend to not be chatty types or forthcoming. This tension is the reality. How do we solve it at Polyface? One way is to have a short meeting once a week to give a broad overview of the big upcoming projects, kind of a flow of the week. Using a whiteboard, we put down leader's names by a project and the number of interns appropriate for that job. Interns then sign up — learning, of course, that the whole thing can be thrown to the wind at a moment's notice.

"The cows are out!" That announcement usually suspends all plans.

In general, the mentor has to think for everyone. That's not a slight against interns. It just means that with more people, the leader has to juggle more options. The more people, the more options. When I'm in charge, I have to push myself to share what I'm thinking. Interns need to strike a balance between showing that they want to know the big picture without pestering or sounding like they don't trust the farmer to lead wisely. I don't have all the answers, but I'm sharing from the heart the real day-to-day land mines in this intricate and intimate dance called farm interning.

Us Against the List

How can mentors close the gap between their heads full of information, the demands of leadership, and the awkwardness of being a teacher when they're unaccustomed to the role? How do you keep those interns engaged, busy, and learning? Let me suggest three things we've found at Polyface that help us leverage intern labor.

First, make lists. Every mentor needs to be a list maker. Lists work. Teresa is a fantastic list maker, and she taught me how to do it. I don't do it much any more because I just do what Daniel tells me to do. He keeps the lists. Rachel has lists all over the place. When I make lists, if I do something that wasn't on the list, I write it on just so I can cross it off and see the accomplishment in black and white.

People who get a lot done make lists. That's what daily planners are all about. At Polyface, we make to-do lists for the year during strategic annual planning sessions. Then we make shorter-term lists, like what must be done this month. In truth, those long-term lists are what any thinking farmer carries around in his head.

The real list — the one that best leverages idle labor — is the filler list. That's a list of short jobs, most of which don't require supervision. Those are the things we see during the day that ought to be done, but they aren't immediately pressing so we dismiss them for another time. Then we forget about them until we go by that area again, and remember that we should have done one of those little tasks.

Most of the things on the filler list are either ongoing projects that can be interrupted easily, or they are self-contained jobs that will take less than half an hour. They fill in those times between breakfast and the start of the day's big project. If eggs come in especially clean and get finished 45 minutes before supper, these are jobs that can leverage that time.

Here's a sample of that list:
- Chop thistles behind the equipment shed
- Replace light bulb in the walk-in freezer
- Grease PTO bearings in the feed buggies
- Pull weeds from the green bean bed
- Sucker tomatoes in hoop house number two
- Mix chain saw gas
- Roll up cross fence in flat field
- Fix small leak in water line by the beehives
- Take wood scraps from the shop to the burn pile
- Straighten electric fence stakes by the barrel
- Fix perch board on the nest box in front Eggmobile
- Set up pig training wire for the new pigs
- Cut out multiflora rose down by the river
- Hang hoses on the sling

You get the picture. This list, which sometimes can be made brand-new every day, keeps people from frittering away valuable minutes between big tasks. It's the kind of stuff that you pound your head trying to remember that you thought of a couple of hours ago, but can't remember right now. This is the way to remember it. Often we compile this list with all the interns present so it stimulates them to think about what they've seen that needs correcting or doing.

As each of these items gets crossed off, the team's sense of accomplishment is palpable. Everyone gets in the spirit and soon the list becomes our shared competitor: us against the list. Who will win? That's a great spirit, and it builds team cohesiveness, encouraging a sense of urgency and awareness. If everyone helps make the list, it builds buy-in and ownership, which in turn keeps people from feeling like they're just following orders. Any time mentors and interns can sit down together and formulate the orders, the whole project goes more smoothly.

Fearless Leaders

The second technique is to have an apprentice manager. Certainly for very small operations, this isn't necessary. But many farms use half a dozen or more interns, and at that point the farmer's administrative desk work increases just to keep things going. This can hamper the process of leveraging the labor. If the interns don't know what to do, they stand around while the mentor handles administrative tasks.

Interns standing around is not a good thing. It stimulates foolishness. Interns hang out enough in the off hours; they certainly don't need to waste valuable daylight and prime working hours goofing off. My sense is that once a farm goes past three interns, most farmers need a lieutenant. That can be an older child, an intern manager, or an apprentice who already went through the internship program. I'll delve into this a bit later when I go through the inner workings of the Polyface program, but at this point I want to make the case for the ratio of leader to followers.

While some outfits may have too many chiefs and not enough Indians, the more common problem is too many Indians and not enough chiefs. That's a sure recipe for wasted time. In my experience, all interns are ready to work; they just don't know what to do or how to do it. They need leadership. In the early days, when we started, we didn't need a lieutenant.

But at about three interns, that's a lot of workers. They need supervision. The whole point of this, remember, is an intimate working-alongside environment. It's not a sink-or-swim kind of deal where the farmer sets up the interns for failure. The whole idea is to minimize mistakes and maximize correct procedures, which requires high personal involvement on the part of both parties.

Fortunately, on our farm, Daniel stepped into that lieutenant role very early — arguably, perhaps too early. When you have a fifteen-year-old in charge of a couple of twenty-year-olds, it can set up some dicey who's-the-boss issues. Fortunately, that period of time didn't last long. But I definitely had to step in and explain the chain of command more than once. "When I'm gone, Daniel is in charge. Period." That way, if a wrong decision occurred, it protected the apprentices from having to take responsibility. Someone has to be the boss. You can't farm by committee.

Some ten years ago we carved out an apprentice manager position, and have filled it more or less ever since. That occurred once we created the internship program, which added six summer people to our already two apprentices. The crew of eight was more than Daniel could handle with me being gone a fair amount. The apprentice manager position has been a life saver. That way if Daniel and I are both tied up for a couple of hours doing administrative work, the apprentice manager can go ahead and get things started.

From an economic perspective, the apprentice manager salary, which is substantial, ensures capturing twice as much labor out of the interns. This means that the interns get more experience under their belts faster, and it means we leverage the small army of labor. If you say the intern manager, a full time position, costs $35,000 a year but yields an extra thirty percent intern labor efficiency, it certainly more than pays for itself.

All of our intern managers have been cream-of-the-crop former apprentices. By having gone through the program, they understand the dynamics of the internship process. To have them complementing our weaknesses is a blessing. All apprentice managers we've had, including our current one, Eric, would be welcome to make a career here. Maybe that will happen. Each of them has blossomed as a leader during their tenures. But I don't know if we'll ever be able to keep one long term — they seem to go off and start their own farms. Oh well, I guess we could lose them to worse things. Ha!

Good Eatin'

The third leveraging technique is to eat one meal together each day. Depending on the type of farm and the preference of the farm family, it doesn't matter which meal it is. Eating together all the time is far too invasive and leads to resentment. All of us need alone time. But one meal is just right for down-processing, increasing familiarity, and morale-building.

When we started the apprentice program with two guys, both of our children were still living with us and with only two interns it was a reasonable thing to add them to our supper, which is our biggest meal of the day. Teresa and I have never done lunch except as a quick in-and-out — it takes too much time to prepare and clean up if you do an honest-to-goodness family-style lunch. My favorite meal of the day is breakfast because I've already been out doing chores for an hour or so and I come in hungry.

Our family has always had a sit-down, cooked-from-scratch, evening meal — Teresa's cooking is legendary. The evening meal offered the opportunity to relax and enjoy each other in a non-work environment. It's where we adopted — emotionally — our apprentices. We've enjoyed lots of fun around that table, let me tell you. Besides, you can't expect the interns to work a full day and then prepare their own evening meal. By offering that, they're happy to work right up until supper time along with the mentor.

We've never done any formal meals on Sunday. Everybody takes a break on Sunday. For years, then, we had the two apprentices in Monday through Saturday for supper, which we serve at 6 p.m. That's when the outside work day officially ends, unless we have something critical we need to do afterwards. Of course you know the old saying, "The farmer works from sun to sun, but the farmer's wife is never done." Although Teresa's normally busy in the kitchen doing clean-up after supper — often with apprentice help — she enjoys this time with the young people to build her relationships with them and to stay in the loop with what's going on. And using her woman's intuition, she sometimes picks up on subtle nuances that go right by me. She's saved my bacon more than once, let me tell you.

Once Daniel and Sheri married and moved out of our house, the apprentices were with them two days a week and with Teresa and me four days. That's still the way it is and it works well. Daniel and Sheri have children and since Teresa and I are now empty nesters, we enjoy having the young people in for our evening meals. Teresa enjoys cooking for more people so it utilizes everyone's talents.

Several years ago when we started the internship program, that added six and then eight people to this meal. That was too much. It's like cooking for Christmas festivities every day. That first year, we didn't cook at all for interns; they were on their own. That arrangement seemed awkward to me to part at the end of evening chores and leave the interns to go fend for themselves. They had worked hard all day and now to go fix their own meal seemed inhospitable.

In addition, Teresa felt out of touch with them because she didn't work with them much. In our house, mealtime provided her the intimacy she wanted to be able to contextualize and relate emotionally with my work-related stories of the day. The distance and separation were discomforting to us, but we weren't sure how to fix it.

That fall, I did a presentation in Kentucky and visited a large produce operation that had built a crude kitchen and patio eating area for their Hispanic workers. I don't think the farm had interns, but the intriguing thing was that for very little money, the farm owners had created a communal space for evening meals. Because the farm had a good reputation in the area, local chefs would come out and cook for the family and crew on a rotating basis. For the chefs, it was a way to connect with their farmer. Rather than charging a catering rate for the meal, the chefs saw it as fun and energizing. The farm paid them a token amount to cover travel expenses and some time, so although the farm incurred some expense, it was not outrageous. The chefs prepared the meals with food available from the farm. That turned the meal into a bit of a game for the chefs, to create something from the farm production.

That intrigued me. The courtesy and hospitality, indeed the familial environment engendered by this setup, was palpable and permeated the workplace. Workers were on their own for breakfast and lunch, but the evening meal offered a wind-down fellowship opportunity as a capstone for the day.

The next year, we began a rotation of interns into our evening meals, one or two at a time, so that once a week we enjoyed an evening meal together with each one. That was better, but it still didn't seem like enough. On a separate mission, I had been courting someone to come and take over the Polyface gardens as a new and distinct enterprise. In the early days, Teresa and I had a wonderful garden. But as the farm became more famous and my travel schedule more hectic, I couldn't keep it up like I wanted.

I stewed about how to have my cake and eat it too — how to maintain the performance and educational schedule but still enjoy food from my own garden. Daniel didn't enjoy the garden as much as I did, reasoning that in the time it took to grow a bushel of green beans, he could raise and sell $500 worth of pastured chickens. From a strict business perspective, he was right on.

But I'm old school, too saturated with homesteading magazines, self-reliance, and a longing to be surrounded viscerally with the earth's cornucopia of bounty. I love being able to step out the back door into an object lesson of God's provision and a deep sense of abundance. Scheming for a gardener, I began putting hooks in the water to see if I could attract someone who would layer a produce operation at Polyface as an autonomous entrepreneurial enterprise rather than simply work for us as an employee. Then I coupled the idea of a chef to cook the Monday-Friday evening meals during the four-month intern period with the horticultural idea. I reasoned that there ought to be an income by putting the two opportunities together.

The next year, sure enough, one of our former interns called out of the blue and expressed interest. Dan Solberg looked over the situation and stepped up to the plate. We immediately developed Daniel's basement to accommodate a kitchen, poured concrete and installed roofing underneath his deck, and turned the whole area into a kitchen-patio dining area big enough for about thirty people.

Finally we had a communal evening meal for both Salatin families along with all the apprentices and interns, managers, and guests every Monday through Friday. Team dynamics changed dramatically. Suddenly all of us were together at the end of the day, unwinding and just enjoying each other. Spirits soared, as did loyalty and togetherness.

I began bringing news items or interesting letters that came across my desk. The meals, created from the farm's production, were outstanding. The interns looked forward to meal times instead of dreading having to fix something themselves. They also didn't mind working hard right up until dinner time. We ended up getting an additional hour's work per day out of the team. Teresa and Sheri received a much deserved break from kitchen responsibilities. On Saturday nights during this four-month summer intern period, Teresa and I still have the apprentices in along with the two weekend interns. Daniel and Sheri enjoy a family-only meal on Saturdays.

Teresa feels strongly that if we ask interns to work hard all day, they shouldn't have to face a do-it-yourself meal at the end. This dutiful hospitality in her spirit, of course, endears her to all the young people as "Mom," a label she wears with distinction and honor. Don't let anybody kid you about who is the real power at Polyface. Ha!

These communal meals offer opportunities to discuss the events of the day, to plan tomorrow's activities, and to build deep relationships that carry us through the rough spots. Our per-plate cost runs about ten dollars each, so these aren't cheap meals. But the investment is definitely worth it for the value it brings to the mentoring and interning experiences alike. A daily communal meal offers a way for us as farmer mentors to show our appreciation to these young people, to bring them into our lives, and to emphasize our love for them. This is family.

After Dan's two years, another former intern, Brie Aronson, asked to take over the summer chef responsibilities. The seamless provision has been a tremendous blessing, and she has refined the evening meals with typical feminine touches. She procured table cloths. It's a little thing that adds an ambiance upgrade to what had been a more picnic style decor. She added flowers. She keeps simple cups of flowers on each table. Again, it might not seem like much, but the beauty and thoughtfulness speaks volumes to the interns: You're special here.

But Brie didn't stop there. The kicker was her name rocks. She found small round rocks and painted each person's name on them — including all of us Salatins, the grandchildren, my mother, apprentices, apprentice manager, and interns — nearly twenty-five of us at this writing. To keep cliques from developing — and mealtime is a prime place to notice them — she puts these rocks out Monday through Thursday (Friday is personal choice seating) like nameplates. Each person must find their rock and then sit there.

At the beginning of the season, in her introductory welcome speech, Brie said, "We've placed your names on these rocks to show that each of you is special to us. We're glad you're here, and we're excited to provide you with a place to grow, learn, and spread your wings. You're not just a group; you're a team of individuals that we honor and respect for the gifts and talents you bring to Polyface. So welcome and let's have a great summer."

Amazingly, I didn't know anything about the table cloths, the flowers, or the rocks. Brie made that amazing speech all on her own, without consulting me. I was sitting there in tears, watching this amazing young lady who but two years before had been an intern and now saw herself as a stakeholder, a co-owner of this magic farm. Loyal to the farm, to us Salatins, Brie personified self-empowerment, initiative, creativity, and dignity. In that moment, tears welling in my eyes, looking into the faces of our nine expectant and certainly a bit overwhelmed interns, all the lost tools, broken equipment, relationship heartaches, and countless explanations were forgotten. All I saw was opportunity and optimism. What a joy.

Our culture is starved for human connection and genuine appreciation. Money is good, but deep gratitude and love are better. The evening weekday meals have leveraged the interns more than anything we could have ever envisioned. I guarantee that if we paid them the $1,000 apiece that we spend on their meals rather than providing this evening meal setting, the experience wouldn't be half what it is. What these interns want from us is time and relationship. These meals offer that, and in turn, the interns work their hearts out for us. It's a fair trade.

The rule of sowing and reaping is as real as the law of gravity. If we sow abundance, we will reap abundantly. By investing in these interns economically and emotionally, we enjoy their undying loyalty and effort. We only get back what we deserve. We only deserve what we've given. We only give out of our heart. Open your heart, and these young people will reward you beyond anything you could ever hope or dream.

Chapter 13

This Isn't Working

What happens when things don't work in an internship or apprenticeship?

I wish I could guarantee a formula that would ensure success with every intern. Yet, as often as not, incompatibility originates with the mentor. I don't want anyone reading this to assume that a complete breakdown with an intern is automatically the intern's fault. Quite the contrary, it's probably about half-and-half. That may surprise mentors.

I hope that in this book I've presented loudly and clearly the attitude of love, forgiveness, patience, and essentially *adoption* that must be exhibited by mentors. Those who think interns are just cheap labor, who don't allow them into their homes, who treat them as second-class citizens or as mere employees, rather than as students at our knees, and speak gruffly to them, won't make good mentors. Farmer-mentors who treat their interns poorly will invariably have a higher turnover rate. I've visited some farms where half a dozen interns have been dismissed in a couple of months. That indicates a mentor problem, not an intern problem. Team stability is virtually always created by the coach.

We have a rule at our farm regarding eggs: if you have to wash more than ten percent of them, something is wrong with your management of the laying operation. Eggs should be clean if everything is right. If they tend to be dirty, the problem is always some problem in management. A hen always lays a clean egg. Likewise, the mentor-intern relationship should be clean and easy, not dirty and difficult. If a farm is going through interns like water, something is wrong, and it's not the interns' faults. I've found interns quite forgiving as long as they feel respected, loved, and appreciated by a mentor.

Yet, just as sometimes a mentor cannot cultivate a climate that draws loyalty from interns, sometimes an intern creates such exasperation on the part of the mentor that the relationship becomes untenable. Such a breakdown should be rare, but be assured that as long as anything involves people, it will have problems.

After a couple of bad intern situations hit us at Polyface, we toyed with the idea of a contract. I know of some farms that require a contract, or some other type of written agreement. Although we eventually opted not to go that route, I certainly don't fault people who do. Essentially this agreement, though it has no real legal standing, articulates expectations. I suppose you could create one that does have legal standing, but do you really want to sue an intern? Or sue a mentor? Forget it.

It seems to me that an agreement, to be fair, should articulate responsibilities or expectations on both mentors and interns. How would I feel if an intern presented me with a contract requiring me to provide housing, Internet service, laundry services, food, and labor-rights oriented working conditions? Would I sign such a document? No.

The devil is in the details. What kind of housing? How much Internet service? Unlimited movie downloads? What kind of food? Does food include soda and peanut butter? And what in the world are "labor-rights oriented" working conditions? Does this mean, like the military, that we have to follow every work detail with a trailer-mounted Port-a-Potty?

As we examined the contract idea, we realized that it simply couldn't articulate the nuances of the arrangement at Polyface. If we promised interns a certain time frame for the internship, what if we wanted to terminate them early? How do you articulate just cause in such an instance? If we required interns to work happily, how do we encourage them to be open with us about their thoughts and feelings? By the same token, if we put in the contract that they are to work like dogs and we are the only and final arbiter, that demeans their personhood.

The point we finally came to was that we couldn't codify something as family-oriented as this. I've never been a fan of big contracts anyway, as you will see in later chapters describing our independent Polyface contractor memos of understanding. Ellis, my speaking agent at American Talent Group, endeared himself to me on our introductory phone call when I asked how we should proceed in our arrangement to have him be my representative. I'll never forget his colorful response that, "Oh, we don't need a contract. If you like me, you'll stay with me. If I decide not to like you, I'll quit. If it's working, we'll be in love, and if it's not, we'll part." I grabbed him on the spot and couldn't be happier.

Even a marriage, which I take very seriously, very sacredly, ultimately has nothing to do with a piece of paper or license. In fact, I don't think the government should be involved in marriage at all. Let people make their vows and if they mean to keep them, they'll hang in there. If they don't, all the signatures, licenses, and certificates in the world won't keep them together. In the final analysis, working relationships depend on trust and mutual respect. Absent that, all the paperwork in the world won't guarantee anything. That said, however, I do think written expectations are helpful, and we'll delve into those later.

We concluded that, for us, everything revolves around spirit, attitude, and aptitude, and those can't be narrowly codified. We realized that the more we relied on paper, the more we'd open ourselves up to parsing words and arguing over magnitude of abridgment. Better to just let it be free wheeling.

That's not as bad as it may sound, actually. The intern wants to have a great experience; indeed, deserves to have a great experience. We've never had an intern who didn't want to be successful. Nobody wants to start into something and fail. We know that every intern will try. But what if that's not good enough?

As a mentor, I have a built-in bias toward keeping the intern. Why in the world would I go to all this trouble and effort if I didn't want an intern? I'm predisposed to wanting it to work just as much as the intern.

The fact that the intern can leave at any time — because there is no contract — keeps me from abusing him. That makes me want to work out differences in a respectful way. By the same token, the fact that I can dismiss an intern at any time forces him to try to please me. The shared risk of failure, not to mention the shared opportunity for success, frame the relationship with mutual interests.

Anyone who thinks my refusal to have a contract indicates that I'm just a tyrant swaggering around abusing young people doesn't have a clue about normal business relationships. I don't know any boss who wants disgruntled folks working for him. Certainly some bosses are jerks. But they're the exception, not the rule. And most of us desperately want to please the person we're working for — unless we're jerks. The shared risks and returns tend to equal out after all, contract or no contract.

I suppose you could say that our application question about the applicant's ability to complete the whole time period amounts to an agreement. Nobody gets past the Polyface application questionnaire who fudges on that question. Okay, so maybe that is an agreement, but that's the extent of it. We've found that the shared desire to maintain a happy environment trumps all documents. Instead of stacking our lives with more paperwork, we'd rather put our time and attention on investing in these relationships.

Perhaps the thing to remember, for both mentor and intern, as we enter this discussion, is that we all wrestle with how much frustration is necessary before saying something. This is true in every relationship, including marriages. Every spouse does something that rubs the other one. It could be the proverbial front or back unrolling of the toilet paper or squeeze vs. roll on the toothpaste. It could be a mannerism, habit, whatever. But in love, you disregard it, realizing you're doing things that rub the other person a little wrong. You pick your battles. Who wants to be combative?

Any contention big enough to merit verbalizing already has lots of thought and perception behind it. An outburst doesn't come from nothing. It simply verbalizes what's been steaming for awhile. This is important for both parties to keep in mind. If I verbalized everything I disapproved of in interns, they'd think I was just a complainer. Realize that nothing gets verbalized from a vacuum. Something is there, itching for awhile, or at a magnitude, that triggers verbalizing the contention. Probably most of us err on waiting too long rather than saying something too early. The longer we wait, the longer the issue festers and the more raw it becomes. Saying something gently early is always better than waiting until it develops into a big deal.

In this chapter I'm going to share the good, the bad, and the ugly with real life stories from our Polyface experience. I've picked illustrative situations in the hope that they will help both mentors and interns realize a couple of things. First, that if you have a problem, you aren't the first one to experience it. Second, if you have a problem, it's not the end of the world. It might be the beginning of a beautiful thing. Third, some problems can't be fixed in a time frame fast enough for either party, and in those instances, retreat is better than perpetual argument.

We had an apprentice in the early days, prior to the internship program, who arrived and simply wouldn't talk. I mean, in a week, he said about twenty words. He worked fine and got along with everyone, but he was too quiet. It was eerie. At dinner, we'd all be chattering away about the events of the day or whatever, and he'd eat without ever entering into the conversation.

He seemed happy and content, but the silence was suffocating. We didn't know what he was thinking, or how he was thinking because he offered no feedback at all. It was too weird. We Salatins decided we couldn't handle the silent treatment. I had a sit-down with him and told him that if he didn't start verbally interacting with us within twenty-four hours, we'd have to terminate him.

I didn't know what was going through his mind, but if I were a gambling man, I wouldn't have put a nickel on a conversion quick and dramatic enough for us to want to retain him. Our minds were made up. We had mentally folded. We'd done everything gentle to encourage him to talk more, telling him that we were uncomfortable with his silence. Short of termination, we had communicated as forthrightly as possible that he needed to talk to us. But based on his behavior thus far, we had no reason to believe that this ultimatum would have any additional effect.

To our absolute astonishment, that evening at dinner he chattered like a school boy. In two short hours he went from awkwardly saying nothing to interacting completely normally. You couldn't pry words out of him with a pair of pliers in the afternoon, yet by the evening he was a jovial, appropriately talkative person. I couldn't have been more surprised if you told me the government was going to quit lying tomorrow. It was the most dramatic shift in personality I'd ever witnessed, and it still ranks as the number one conversion in our Polyface program. We kept him and he finished fine.

So let me tell you about another success story. We had an extremely smart, older apprentice, again back in the day before interns, whose questions always had an edge. We call them "questions with a barb." You can ask questions two ways. One assumes a place of complete open interest; the other has an air of distrust or prejudice attached to it. Sometimes this can be as subtle as voice inflection rather than the actual words spoken. Daniel and I were feeling burdened by this spirit, so I had a sit-down with the apprentice and explained the situation.

He was totally unaware of the problem and assured me that what we saw as an arrogance was just a quirk of his personality. In spite of that initial defensiveness about my concern, he quickly changed. The mood of communication eased up immediately and he finished his apprenticeship wonderfully. The lesson learned is that it's worth it to clear the air if you have a concern.

I can't resist — here's one final success story. Every apprenticeship or internship occurs at a unique point in Polyface history. Sometimes it's a special place in our business cycle. Perhaps we've just rented a new farm and have to develop it to receive a controlled grazing program. Maybe we've added a new significant restaurant account that needs special product handling as we pack the delivery truck. Maybe it's the summer we plan to invest in building two new ponds up in the mountain. Sometimes it's a unique and unexpected weather occurrence, such as a flood, hurricane, or in 2012, the derecho that obliterated twenty-five of our pastured poultry shelters, killed a thousand chickens and turkeys, took off a barn roof, and blew over countless trees on power lines and fences. That was memorable.

Daniel and Sheri built their house with our own farm-milled lumber from trees we cut from our forested acreage. But like any owner-built home, it didn't go from footers to finished in a month. They lived in a small recreational trailer for awhile until they could get into the basement. Then they began working upstairs. Our apprentice at the time became increasingly belligerent about helping on the house, which was the special farm project at the time.

"I didn't come to build a house," he complained. So I had a sit-down with him to explain that right now, today, this is the most important job on the farm. We were all hustling to keep up with chores and regular farm work, going to the house project with every spare minute. It was going well, but this apprentice didn't want to help with construction. I told him we'd have to let him go if he didn't have a change of heart within twenty-four hours. I told him we didn't need him.

Remember, if your farm won't survive without the intern help, you're not ready for an internship program. Dependency will make you put up with disrespect, laziness, and incompetence beyond the point of reason. Certain individuals definitely provide something unique to any business, including farms. But the farm must have continuity beyond those individuals if it is to be a resilient and viable business. Ultimately every single person is replaceable. Thinking or planning otherwise compromises your judgment and jeopardizes the business.

Fortunately, this apprentice had a come-to-Jesus moment, got his heart right, and all was well after that. The tipping point was when he realized that we'd rather get rid of him than put up with his whining. He really thought he was too important to dismiss. When he found out otherwise, he sobered up. Those are the success stories.

Termination Scenarios

Unfortunately we have some similar stories that aren't so successful. You never win 'em all.

The first failure occurred early, with a newly married couple, that resulted as much from too many life changing events for them at once — marriage, interning, leaving home — as it did from any weakness in character or attitude. From that experience sprang our bias against couples doing the internship. Too many competing loyalties.

Again, this is not to say we don't allow married people to do the internship; it means we don't allow internships as a couple. The intimate experience in the crucible of daily work and learning militates against a clear chain of command and the undivided loyalty we need for unity on the team. If a wife complains to her husband about her treatment at the hands of another intern, that violates the chain of command we want to occur, which is to come directly to Polyface staff. It's only natural, of course, for her to go to her husband first, and it's only natural for him to want to deal with the issue himself.

The same is true for her if she perceives he picked the short straw on some less-than-enjoyable project. Rather than come to Polyface staff with her complaint, she foments resentment in his spirit, even if there was none to begin with, creating dissension in the ranks. Newly married couples often want to be assigned to projects together, which complicates more free-wheeling manpower allocation. It's an unnecessary tension. As we approached termination, the couple realized what was developing and resigned from the internship rather than being dismissed. This was big of them and eased what would've otherwise been a more painful situation.

One apprentice came to me at about the nine month point in his twelve-month program and announced that, "I think I've learned everything there is to learn here." Without batting an eye, I leveled back, "Well, you'd better leave by tomorrow morning before you forget it all." He was gone quickly.

Another apprentice, prior to the intern program, accused us of destroying his spiritual life. He debated with us about everything and it became intolerable. We parted ways.

Probably the most difficult situation was an apprentice, again early on before the intern program, who meant well but never had any situational awareness. This is a requirement for farming success. Being able to see whether the tractor you're standing behind is going forward or backward is an important thing to perceive. Being able to tell if a gate is open or closed is important. Noticing that water is spewing in a twelve-foot geyser out of a pipe is critical if you're going to succeed in this business.

We even called his parents and asked them if he had learning disabilities. His spirit and attitude were wonderful, but we actually became concerned that he would perish due to lack of awareness. He couldn't follow instructions, couldn't respond to contexts, couldn't make judgments based on changed circumstances. It was both frustrating and frightening. The final straw occurred one day when the herd of three hundred cows got out of their paddock at one of our leased farms where the landlord operates a horse boarding/ training business.

A worker on their horse enterprise there had failed to close an alley gate properly and the curious cows began rubbing on it and popped it open. This opened up the entire horse enterprise to these cows, who quickly poured through the open gate, spreading out across the entire horse area — about 25 acres of paddocks, lanes, outdoor arenas, indoor arenas, equipment storage and tack rooms. The panicked horse staff called us and we went over to extricate the cows.

In short order we gathered the cows into one of the alleys between the horse paddocks. We dispatched the apprentice to the bottom of the lane to open a gate so the cows could loop around and re-enter the field where they were supposed to be. We didn't care if they re-entered the field from where they had exited or from the opposite end, so once we established a direction, we worked them from the opposite end. The apprentice walked off down the lane to open the gate back into the paddock, but in typical fashion didn't lift his head or cock his ear to be aware of what was going on anywhere around him.

Meanwhile, completely unbeknownst to him, the herd, which by now was stirred up a bit by the early morning jaunt through interesting territory, broke into a run — known in herbivore lingo as a stampede — rather than their customary gently-ambling stroll. The whole three hundred of them took off down that lane, a narrow ten-foot alley with a woven wire fence on one side and a board fence on the other side. The apprentice was in this alley on his way to open the gate.

From our vantage point, we could see both the herd and the apprentice. Head down and ambling along, the apprentice was oblivious to the rapidly-closing gap behind him. We screamed, yelled, waved our arms, screamed some more, running as fast as we could to get his attention. At the last second, he heard the screams, lifted his head, and at the very, very last moment leapt up onto the board fence while the whole stampede went by. Of course, he hadn't made it to the gate he was supposed to open back into the paddock in time so the whole herd went thundering past it and we had to restart the whole process.

We actually feared he'd be killed. He would've been killed had he not heard our urgent shouts. He never knew the herd was bearing down on him, even though by the time he heard them they were just a few narrow feet away. That incident was the final straw; we realized we couldn't afford to keep him. It was emotionally wrenching for us because he had a great attitude and wanted to stay. When we told him, he wept — and so did I.

As much as we'd like to think desire can overcome anything, it can't. Sometimes the people and the program are out of sync. When it doesn't fit, it doesn't fit. You can't force it. This is why you dare not go into internship or apprenticeship programs with the idea that your business is dependent on them. That will jaundice your judgment when a difficult decision needs to be made.

A similar situation developed a few years ago with an apprentice who hadn't gone through the internship first, who couldn't progress in learning and mastering the basics that we require for success at Polyface. I suppose if this were a physical problem, it would be classified as failure to thrive. As with any educational trajectory, we start these young people out with the most simple tasks, closely supervised, and as they develop competency we move them up the responsibility ladder to more difficult things.

Interns generally aspire to take on more responsibility. They know when we trust them enough to send them off by themselves to do a particular project. The interns yearn to receive this kind of trust from us. Although we don't necessarily accompany some of these responsibility promotions with cymbals and fanfare, interns know when they've crossed that threshold of trust and additional autonomy. This is what we strive for.

Now, every intern will make boneheaded decisions and botch up things — especially on the front-end of the program. We expect it and that's why part of minimizing our risk with novices is keeping the jobs relatively brainless and heavily supervised. In pet parlance, this would be called a short leash.

The goal is to gradually develop reasoning and observational skills so that more complexity, autonomy, and responsibility can increasingly be entrusted to the intern. We have a phrase around Polyface: "Don't tell me what went wrong. Tell me how you fixed it." (America's current cultural victim mentality is sometimes hard to break. Whether it's people sitting around their homes after a natural disaster waiting for the government relief agencies to do all the lifting or an intern who walks away from a problem, the underlying dependency mentality is the same. Anyone who plans to succeed in life has to break that dependency.)

Daniel says the biggest weak link among all interns is creativity. You'd think in a day and age of video games and fantasy play creativity wouldn't be in such short supply, but it is. It probably always has been. But at Polyface, we can't afford people who aren't creative. We need creativity every moment of the day, both for expected and unexpected problems. I tell the interns they should walk around all day with their head up and eyes wide open looking for problems — a leaky water line, a cow out of place, a bucket of water on a chicken shelter that's six inches different than the other twenty. All the shelters are the same, same birds, same day; the water level should be pretty close to the same also. If it's not, it indicates something wrong. Any aberrant thing should attract your attention . . . BECAUSE you're looking for it.

At Polyface we prize creativity and attention to detail. A lot of people cavalierly dismiss details with phrases like, "don't sweat the details." Let me tell you something: you'd better be glad somebody does sweat the details. Do you want to take a flight on an airplane where the mechanics didn't sweat the details? Do you want to buy a computer built by someone who didn't sweat the details? Interns who thrive are those who understand the importance of creativity and sweating the details. Writing this book, I'm sweating the details so you don't get assaulted with misspellings and typos. I know some will exist, but I sure try to minimize them. Sorry for the ones that slipped.

Now back to the apprentice who failed to thrive. If something broke, he didn't try to figure out how to do the job another way, or how to fix it. He just walked away and reported the problem. What really catapults an intern into management positions is the ability to pursue options, creatively, until he exhausts all options. The thinking process is truly what separates the stars from the also-rans.

After even several months' time, this apprentice couldn't be trusted to do anything. Again, this was gut-wrenching for us because he wanted to finish the apprenticeship. He wasn't belligerent or complaining. But every job, even the most menial, required someone to go and check behind him. It became intolerable and we had to let him go. This was the experience that finally caused us to quit taking apprentices without vetting them through the internship program first. Hopefully this will increase our batting average and reduce "I wish I'd done this differently."

We know we're dealing with hopes and dreams when young would-be farmers come to us as interns and apprentices. Raw emotions are bound up in these aspirations, with life-changing assessments and trajectories at stake. The flip side is also true: internships are not about fun and games. They're not a substitute for excellence. Internships put the practical flesh on the bones of what you think you know and they require the same level of dedication and performance excellence as anything else in life. They're not just a walk in the park, but an arduous striving to master skills. As difficult as it is to terminate an internship, keeping an intern who isn't working out is far more difficult.

When I was in college, I remember thinking how unfair the sophomore platforms (performance tests) were. In certain disciplines, like speech, music, and theater — usually some sort of performing arts — sophomores had to present a platform at the end of their second year. A panel of faculty judges decided whether the student had enough talent to stay in that major. I remember friends who failed. They were devastated.

Always quick to defend the underdog, I often took their side. How dare a panel of judges decide that a piano music major couldn't proceed in that major? It seemed tyrannical to me. What if that piano player decided to play differently, outside the norms of

211

conventional pedagogy? Was there room for new ideas? The world is full of geniuses who failed such platforms. Einstein flunked out of school. He failed to thrive within the school paradigm.

But I've come to appreciate these platforms in a broader context. The school is trying to turn out representatives who will carry the brand name into the culture. The school doesn't want its reputation besmirched by graduates who fail to thrive. Just because a student doesn't fit there doesn't mean she won't fit somewhere else. Or perhaps she can change her major to something more fitting, which is often the case.

At Polyface, we have very real work to do. We also have an image, a brand name, a public face to protect. This is why we care about grooming, appearance, and language use. Failure to thrive with us doesn't mean the end of a career in farming or farm-related businesses. So far, our selection assessment process has been pretty effective for us and for our interns and apprentices. Termination is never fun, but it needs to always be an option for both parties in order to end the stress of a poor fit. If a shoe is wearing bunions on your toes, it's better to get rid of the shoes than prune the toes. Termination is as much a beginning as it is an ending. How many entrepreneurs started a business as a result of job termination? Many will enthusiastically say that termination was the best thing that ever happened to them.

When things degenerate to the point where termination is necessary, the difficult emotional tension will always give way to relief. Firing an intern is always a last resort, but when it's necessary, it's a necessarily good thing. It happens.

The great thing about interns that work out is that they can become members of your family. But the affection that develops between farmer and intern can make it extra difficult when the time comes to recognize that the relationship just isn't working out.

Ending an internship early can be your last lesson to a young person, a lesson in resilience and the value of moving on to something that works better for him. As for you, the cock will crow in the morning and tomorrow will be another day of hard work in the fresh air.

Future Farming Success

Chapter 14

Accessing Land

Perhaps nothing occupies the mind of an aspiring farmer more than the question of access to land. Amazingly, nothing occupies the mind of an established farmer more than what to do with the land — not just today, but in perpetuity. Here we have these two people, sometimes living on the same road, agonizing over the opposite ends of the same problem. It's quite a conundrum.

I've often thought that the people who were the least creative and least deserved to own the land actually owned most of it. The underutilized and eroding land more often than not can't be accessed by creative young energy. But creative young energy is not enough, as I'm hoping this book explains. It's not enough to be excited; the excitement needs to be guided by wisdom and skill.

All that being said, I'll take a young, enthusiastic novice over a pessimistic hermit curmudgeon any time of day. Be that as it may, a farmer can't do much without land. It kind of goes with the territory. But I hope we've established the fact that it might not take as much land as you think.

Will Allen, urban farming guru and founder of Growing Power, says that in his most developed models he can get a million dollars worth of production income out of a single acre. Mind you, that's an all-season acre under hoop houses utilizing all the vertical space with multi-speciation and complementary enterprises. But think about one million per acre! That's serious; I don't care what kind of infrastructure it takes.

By the way, Will operates an internship program too. He told me that his satellite enterprises, which he now has scattered in cities throughout the U.S., are much more successful when operated by intern graduates. In some cases, community groups embrace Growing Power's ideas and try to duplicate Will's genius on their own or have him consult for them. These flounder. Will said his intern grads don't flounder in these satellite operations. Just another testament to the magic of interning and mentoring.

As we begin this discussion, the main thing we need to remember is that you don't have to own land; all you need is a piece of land to control. When my parents bought our farm back in 1961, the land was $90 per acre and 500-pound calves sold for $30 a hundredweight, a ratio of 3:1. Today, the land is worth $6,000 per acre and those calves sell for $120 a hundredweight, a ratio of 50:1. If you don't know anything about agricultural economics, read those two sentences again. Let the ratios sink in. This astronomical shift in land price relative to production value is why acquiring land without an accumulated nest egg is nearly impossible. By the way, that shift in land value versus production value was not accompanied with an increase in sunshine, water, or soil.

What this means is that most farmland being purchased today is not being purchased by farmers. It's being purchased by non-farmers. Of course, the non-farmer category is huge, including not only the wealthy stock trader but also land trusts and preservation organizations. In some cases, even cities are buying up defunct farms to preserve green space. The over-arching principle is this: owning land and running a farm are two separate businesses. To make a farm capitalize the land it occupies is an unfair burden based on an incorrect understanding of today's farm economics.

As I've pointed out elsewhere, plenty of land is available. And to be sure, even most privately owned farmland could be significantly more productive. Allan Nation, editor of *Stockman Grass Farmer* magazine, says that we may be entering a farm model that looks more European. Old money owns the land as a defensive economic posture, but it is farmed by younger people, and younger money, as an offensive posture. One group is trying to hold onto wealth and the other is trying to acquire wealth.

Acquire wealth through farming, you say? Absolutely. The fact that universally people believe "there ain't no money in farmin'" simply creates an open door for anyone who figures out how to do it. The universe of truly profitable farmers is small enough to create unprecedented opportunities.

How so? Land leases are based on this low profitability assumption. Owning and renting are at two ends of a teeter-totter. As one goes up, the other comes down. We see the same thing play out in the housing market as these two options vie for ascendancy. Plenty of financial gurus have their formulas and benchmarks to assess whether it's better to buy a house or rent one. Generally, as houses become overpriced, renting becomes a better option. When the housing market plummets and nobody is buying, rent goes up.

That is exactly what has happened with farmland. With this unprecedented disparity between sale price and production value, the profitability of farming, both real and imagined, is low. Owning a piece of property requires money if it has any infrastructure you want to keep up. Fences, sheds, houses — these take money to maintain. A farm ceasing to produce income quickly becomes a wilderness area. Stewardship requires effort, and somebody has to pay for that effort. Some income is required just to pay the property taxes.

Farm land value today generally has nothing to do with farming. It has to do with perceptions and viewscapes. A rock pile with a pretty view sells for more than excellent bottom land protected by hills and creeks. This creates a real hardship for scenic hardscrabble farms suddenly discovered by wealthy folks. All it takes is one attractive farm to be sold in a community, and suddenly the tax assessment skyrockets. This has created real financial hardship on the farmers who don't want to sell but suffer an escalating tax burden. I like the California rule that no property taxes on a given parcel can go up more than one percent per year, regardless of anything. That's a good policy. Property taxes should not throw people off their land, period.

Leasing

At any rate, farmland leasing is based on the real economics of farming. Since farming, for the most part, is unprofitable, leases are amazingly disparate relative to purchase prices. In our area, good pasture land, for example, sells for $7,000 per acre. The average lease is about $35 per acre per year. This is enough to pay the property taxes and build some new boundary fence.

It certainly won't make a landowner rich, but landowners aren't trying to get rich. They don't own the land to make money — unless they're speculators or traders. They own land for a host of reasons:

1. I always wanted a piece of land to call my own.
2. My fondest memories from childhood are of my grandparents' farm, and I want to resurrect those fond memories.
3. This farm has been in our family for three generations and I don't intend to be the one to lose it.
4. Farms are romantic.
5. I needed a quiet, private place to hunt and fish, so I bought a farm.
6. I want a place to raise my children where they can run and play and learn how to work.
7. I wanted a business my family could do together, and farming seemed the best way to do that.
8. The country's going broke and barbaric; I need a safe haven to flee to when everything collapses.
9. I like to hear roosters crow in the morning.
10. I need a tax shelter and this is a fun one.
11. I wanted a milk cow.
12. I want an exclusive place to entertain my friends, where we can barbecue looking out across an expansive pastoral setting and they can covet my spread. Gloat, gloat, gloat.
13. Farming is American, and I'm an American!
14. I want to heal some land in a visceral way.
15. Growing things just runs through me; I've got to try.
16. Real men farm.
17. Real women farm.

You get the picture. None of these has much to do with operating a business. The motives have to do with yearning, dreams, fantasies and obligations. That's why Dave Pratt of the Ranching for Profit Schools advises even the farmers who own their own land to charge themselves a going lease, in order to keep the business figures honest. Today, precious few people buy farms to make money. That's decidedly different than it was a mere fifty years ago.

The point is that a million dollar piece of land generally rents for less than five thousand dollars per year. You could rent that farm for two hundred years, at the going rate, before you paid what it would cost to own it. Given that economic difference, you can see quickly that leasing is a better option. Greg Judy, whose first book, *No Risk Ranching*, explains this lease-own disparity, enjoys living in an area where some landowners pay him to lease their farms. In such a farm-depressed area land can be leased for nothing.

One of our former interns returned to his home in upstate New York and within a month was offered a total of a thousand acres from three land owners for free. "Just please come and do something with it," they begged. So far, in our two decades of mentoring, we've never had a Polyface internship graduate unable to find land. As aging farmers go out of business or die without children to take over, more and more land becomes available. This trend will likely escalate over the next twenty years.

That bodes well for young people entering farming. Europe developed a system called tenure, commonly known as ninety-nine year leases. In fact, some farmland preservation groups in the U.S. are now holding seminars to acquaint people with tenure. Although modern Americans tend to think of this term applying only to protected and permanent college teaching positions, it's actually rooted in stable land management.

The obvious concern for farmers tied only to a lease is its temporary and tenuous nature. What happens if the land owner decides to sell the land, or change the lease? The *catch-22* is that land owners want freedom to do what they want with their property. However, that freedom creates instability for the farmer who leases that property. All farmers negotiate for as long a lease as possible.

In the end, you can't have it both ways. The landowner must give up some freedom and the leasing farmer (land-manager) needs to handle some risk. Obviously going to a ninety-nine year lease takes away the landowner's freedom completely and gives the land-manager farmer ultimate stability. It seems to me that people willing to place their land in conservation easements or similar land preservation covenants would be open to long-term leases.

The problem is that most land preservation covenants aren't written to accommodate future innovation. Most covenants preclude things like day camps, agri-tourism, milk bottling, processing, commercial kitchens, on-farm stores and the like. They preserve only the ability to produce commodities, not the ability to process, distribute and market those items. Some farmers who signed up for these covenants years ago are now suing the trust-holders for the right to bottle milk or mill lumber on these protected farms.

It's as if the protection extends even to protecting them from profitability. The fact is that times change, and it's extremely difficult to codify land use appropriate to tomorrow's climate. That unknown prospective climate includes weather, economics, and people resources. What if a future farmer doesn't want to just grow potatoes for McDonald's french fries, but would rather turn those potatoes into spud-sculpture through an on-farm craft store?

This is a difficult issue in the land preservation movement and it's why I encourage farmers to not jump into easements and covenants. Too often the innovations necessary to preserve farm profitability don't fit within the narrow bucolic pastoral assumptions of the covenant holders. These covenants gained wide acceptance during the farm crisis of the 1980s coupled with rapid-fire urbanization, especially the land-gobbling suburbs and related buildings. This coincided with an unprecedented industrialization in agriculture, including the phenomenon of the Concentrated Animal Feeding Operation (CAFO).

In virtually all quarters, the society was stratifying and specializing. Today, the information economy and regenerative economy are working together to re-localize, down-size, and re-structure. The urban center "donuts," as land planners call them, now enjoy revitalized core economies. Young people aren't buying cars. Walkable communities integrate diversity. Part of that is the trend for farms to re-house small-scale manufacturing, processing, and distribution to serve these connected socio-economic sectors.

That means functional farms in 2015 won't look like the sequestered farms of the 1970s that these covenants try to preserve. Freezing a sequestered system isn't the way to create an integrated future.

Long-term understanding of land use, or more specifically, appropriate farm use, is almost impossible. How would you like to tell Google what it should look like in fifty years? Would anyone suggest freezing Google's looks to today's image? Anyone who knows anything would laugh at the notion, realizing that Google will morph, move, and re-invent itself in the rapidly changing climate of innovation. And yet that's exactly what the easement holders do to farms. It's a form of bondage, often tying the hands of innovative farmers.

The same pitfalls, of course, befall privately arranged long-term leases. Landowners have a certain look in mind, but often have a hard time keeping up with farming's innovations or new realities, such as changing from a 3:1 land/production ratio to a 50:1 ratio. Or like changing from a specialized to a holistic society. Or moving from global dependency to local sufficiency.

Recently I had the privilege of spending nearly a week on an aristocratic estate in England dating back to the early 1400s. Comprising 17,000 acres and 325 buildings, the estate leases 6,000 of its acres to some 14 farmers in European-style decades-long tenurage arrangements. The owners cannot sell any of the buildings or land. They cannot demand their tenants to quit using chemicals or begin soil-building techniques. The tension in these arrangements is palpable and gave me a renewed appreciation for America's promise as a nation of yeomen each owning at least something.

Ultimately, legal documents aren't as efficient as the integrity and good faith of the parties involved. That's why most long-term leases occur after a couple of short-term rotations. Both parties want a track record before signing on for a long-term deal. Unexpected problems can occur on both sides: leasing farmers can turn out to be deadbeats or pillagers, and landowners can turn out to be arrogant tyrants. In our twenty years of leasing farms, we've only terminated one lease. I likened it to a divorce because when you get onto one of these places, you develop a real love for the land. You learn where the wet spots are, where the dry spots are, and you begin developing it with water and fencing. It doesn't take long to fall in love with the land.

Leasing Relationships

That termination experience made us acutely aware of the danger signs we should have seen prior to signing the lease. Unfortunately, it took us a few years to see what we'd gotten into and eventually walk away. Based on that experience, here are the red flags:

1. *Beware of multiple masters.* This farm was owned by a family trust comprised of three middle-aged siblings and their ailing mother. She passed away in our first year of the lease. At the outset, we didn't establish the need for one point person. The mother was the de-facto boss, but as her health deteriorated, a clear leader among the three siblings didn't emerge.

Consequently, we found ourselves serving four masters and that soon changed to three — which is two landlords too many. You can't serve multiple masters. Each has a pet project, or a landscape plan, or stipulations. It's maddening. The siblings disagreed about the farm's future direction. Each had individual goals for the farm, hamstringing our plans in the process. Trying to please everyone, we ended up pleasing no one. If a farm has multiple ownership, any lease needs to specify one and only one contact person, up front. If that can't be done, walk away. No man can serve two masters.

2. *Recognize a non-magnanimous demeanor.* By this I mean that from the outset, this family micro-analyzed every last thing to the point of stinginess. This twig could be moved; that one couldn't be. This building was half ours and half theirs — down to a chalk line on the concrete floor.

The acreage changed several times in the first season as they poured over aerial maps and old crop subsidy records. One acre more, one acre less. The ten thousand dollar lease was not nearly as important to them as whether it was \$9,985 or \$10,020. The list of rules should have tipped us off, but we initially thought this simply indicated a meticulous nature and clarity of thought. It did not. It indicated a miserly spirit, a closed hand. A lease needs some wiggle room for the unexpected.

For example, when the fierce and dramatic derecho wind storm hit in late June of 2012 and knocked out power — and our water pump — for a week, we ran the cattle herd to a stream for water. They had to drink, right? But the landowners had a hissy fit. Interestingly, the previous renter had run his cattle in the streams routinely, but we fenced them out since we had a policy of keeping the cattle out of riparian areas. Breaking with this policy for a few days during this unprecedented crisis garnered no sympathy from the landowners. From their response, you'd have thought we ran the cows to the creek just to be spiteful, as if breaking with our policy indicated some sort of dark malevolence or trickery. This is what happens when the landlord's demeanor is excessively persnickety. It also happens when the landlord has no clue about the realities of farming.

3. *Guard yourself against a belief in non-participatory environmentalism.* I like the environment, but I don't think human abandonment is the only way to preserve it. These landlords had a virulent no-cut policy on trees. It went beyond that, though, to a no-touch policy. Even running a water line became contentious if we knocked over a tree seedling.

Never mind that the six-hundred acres of forest needed thinning and were full of dead and dying specimens. Good environmentalist landlords enjoy hunting and fishing, cutting trees, building fences, and getting out there and participating in nature viscerally. These owners didn't hunt or anything else. They viewed the land as a big park for swimming, hiking, and singing "Kumbaya" around a driftwood campfire. We essentially became park managers instead of farmers.

So if you see radical environmentalist magazines lying around a potential landlord's house, make sure you diligently ask questions about periodic landscape disturbance and interactive relationships between humans and the environment. If your impression is that the landowner thinks that human breath taints the environment, get out of Dodge quick.

4. *Tune in to any hyper-attachment to nostalgia.* Failing to see this problem was perhaps our biggest blind spot in the original negotiations. The farmstead had numerous buildings. Seriously, the first time we visited we expected to see John Wayne come strolling out of one — it was truly that rustic and Wild West looking! But early on, the landowners created odd touch-embargoes on most of the buildings.

The use-prohibition meant we couldn't open a door and step inside — these were sacred museums in honor of grandfather. These fifties-aged siblings remembered coming to the farm and doing something with grandfather. To darken the door of those buildings, especially with strangers, was tantamount to desecration in their eyes.

Over the course of the four years we leased this farm, these museums became more and more contentious. We had to fence them out so a cow wouldn't rub on them. We couldn't store mineral, or even a thistle hoe, in them. As they began falling apart, they littered the farmstead, but these were their sacred relics of the past and had to be left as-is, declining quality and all. It was over the top. As the relationship deteriorated, they demanded that we replace roofs and doors that had blown off or fallen off. Remember, these were buildings we couldn't use or touch.

The way to keep things up is to use them. Vacancies encourage deterioration. I like shrines and I like farms, but it's hard to make a living on a shrine-farm. The land and buildings need to change with the times, to be used and retrofitted. That's the lifeblood of viable upkeep.

The good part about leaving that place was that it proved the wisdom of the lessons I'm teaching in this book: what works in farming today is portable infrastructure, multiple enterprises, and direct marketing. Because of these principles, the high income we could achieve on the rental farm didn't require that much infrastructure, so it was relatively easy to extract when we made the decision to exit. All of the infrastructure went on a couple of lowboy trailers and was eventually assimilated onto other farms. This cut our losses dramatically.

Elements of a Good Lease

What should a lease look like? We have a whole range of agreements. The simplest is a hand shake — nothing written. The longest is six pages of legalese. One of our landlords has three attorneys in the extended family: guess which lease is six pages long? In the end, the whole arrangement hinges on trust, and you can't codify trust. I like written leases, but if a landlord trusts you enough to shake on it, that works for me, too. Even a casual lease, though, requires proof of liability insurance because most homeowner's insurance underwriters require it.

The basic components of a farm lease are as follows:

1. *Identify the leasor (land owner) and leasee (farmer) and define the property in question by address* (Twenty-Three North XYZ Road) or by geographical description (lying on the west side of XYZ road approximately .3 mile south of the ABC road intersection).

2. *Set the lease length and compensation amounts.* Usually this section will include a price per year or a price per acre as well as any extras, like electricity to run a water pump. It's important to separate these for tax purposes. Rent is unearned income and subject to a 1099 tax form; expense remuneration is not.

3. *Set the renewal terms,* such as how many days prior to lease termination the re-up negotiations must occur.

4. *Articulate responsibilities that fall to the landowner,* such as boundary-fence building. This section may include prohibitions, such as chemical fertilizer, pesticides, hunting, or use for anything other than necessary farming endeavors (i.e. the landowner can't bring a bunch of friends over for a picnic and bonfire on the back forty while the herd is in that paddock).

5. *Articulate responsibilities expected of the farmer,* such as fence maintenance. Often in this section we also include a description of our farming practices and what the land owner may see from

time to time — tall grass stockpiled for future grazing, for example. Because of our grazing management, the fields don't look like golf courses. In late summer or fall, they often look unkempt since we let the forage grow to its full expression. We want to be clear about that in advance.

6. *Articulate a chain of command*, if that's applicable. Because former interns acting as independent contractors actually run day-to-day operations on most of our leased farms, landowners often think that if they communicate with our managers, they've communicated with us. Not so. Although these managers may be the people the landowner sees routinely, the deal is with Polyface and ultimately it's our responsibility to make the landlord happy.

7. *Know the termination requirements*. This lays out how either party can get out of the deal and usually contains some sort of time frame, like ninety days written notice.

8. *Define a disagreement protocol*. At Polyface, we will not sign a lease that allows legal recourse to settle disputes. Every single one of our leases has a litigation prohibition and only allows for binding arbitration wherein we choose a person, the landlord chooses a person, and then those two pick another mutually agreeable individual to act as a three-person jury to hear the dispute. For us, legal action is an absolute deal breaker. Addressing that up front helps us smoke out the potentially litigious nature of the landlord from the get-go. We want everyone we're working with to abhor litigation as much as we do. You'll see the same prohibition in our memorandums of understanding with collaborators.

No matter how meticulous you may try to be in laying out all expectations and contingencies, things will come up. A tree falls across the fence. Three inches of rain fall the night before you have to move a herd of cattle out of the farm and you can't wait because you've already lined up three cattle haulers to arrive at nine a.m. The heavy trailers make ruts in the lane — oops.

The landlord has friends over and one of them drives through the electric fence, letting all the cows out. These things happen, but with shared vision and trust, you can get through them. We've found that every landlord has an itch. If we can find out what that is and scratch it, we get along very well. By the same token, every landlord has a soft spot — stay away from that. It may be letting a particular invasive species grow in the field or fencerow. While that may not be an issue for you, if it is to the landlord--a soft spot--then it's a big deal. It may be how fast you go on the farm lane. Stirring up dust. These soft spots make or break trust. The sooner you can determine these personal wants and taboos, the better.

A load of gravel hauled in and spread on the lane can do wonders for goodwill. If you mess something up, be the first to offer a timely fix. Don't wait until the landlord sees it and calls you. On that note, we spend a great deal of time communicating with our landlords. One thing all landlords share is a distaste for surprises. They like to know what's going on.

You may think clearing out a patch of briars and weeds wouldn't be a great surprise. Yet we did that one time and the landlord accused us of taking away the flowers his children enjoyed — on the multiflora rose bushes, for crying out loud. Landowners are wary of anything different. Catastrophic changes make them feel crowded, like you're taking over and pushing them aside.

You never want them to feel like they've been poor stewards. Even if you're taking over from the most despicable farmer imaginable, never talk down the former renter. Even in the worst circumstances, the landowner-tenant relationship is intimate. Even if it was a bad situation, don't assume they weren't friends. And don't assume the landlord terminated the arrangement happily. The grief over the termination is shared by both parties, just as in a divorce. Don't think you can exploit the separation; just appreciate that it gave you an opportunity and then try to discern the salient points of the disagreement so you don't end up terminated too.

If our landlords don't live on the property, we take pictures during the season and email them explanations so they can enjoy seeing what's going on. We let them know when cows are coming and when they're exiting. We ask permission for everything, even if it's to cut a tree that fell down in a windstorm. We lease one farm that has two fallen-over trees out in the middle of a field. The cows rub on them and muck up around the area. No grass grows under and around them and they're unsightly, with their root balls tipped up. They're bad for cattle hygiene and create a conducive incubation site for parasites. But when we asked to cut them and clean the area, the landlord objected: "Those are like sculptures in the field and are beautiful in silhouette against the setting sun." Oh my. So much for the way I look at things. My eyesore is their art piece.

But that's the way things are. It's not right or wrong, it just is. The relationship between landlord and farmer is much like a marriage. To have a healthy and continuing relationship requires much communication, sacrifice, and attention to detail.

As a farmer who has rented numerous pieces of property, I'd say the most common sticky point is being assured that I have the authority to do what I need to do on this particular piece of land to run a successful farm. Much of the time, if not most of the time, even the best intentioned landlords don't understand farming and what it takes to be successful. The majority have never made a living from the farm. Of the current nine parcels that we rent, only two landlords have ever attempted to make a living from their land. This may actually be a higher percentage than most parts of the country, since our area is quite traditional and still enjoys an agriculture-based economy.

Tying the hands of farmers is especially problematic on land owned by preservation groups. Not only are they bound by an often out-dated and unrealistic conservation easement, but the easement enforcers have their own personal assumptions about what *is* and *is not* acceptable. By the time the poor farmer wades through the written material and then gets permission from the trustees of the covenant, he's often wiped out emotionally and economically. I've seen way too many honeymoon land-marriages dissolve in acrimony within a year because the farmer's hands were tied.

This is why I like to express, both verbally and in writing, exactly what I plan to do with the property. Remember that too much change too fast alarms people. You don't have to put all your cards on the table by stating the most extreme land use possibilities thirty years hence. Keep those for a later day. But you do need to ascertain opposition that would deny you the ability to do the most basic changes needed to be profitable.

Without a doubt, the most common attitudinal hurdle in all these arrangements is the assumption that the landowner is a nobleman and the farmer is a peasant. I assure you that today, even though this is my ninth book and I routinely speak to audiences all over the world, I'm still considered a farm hick. I bristled when I went on the Bloomberg Business TV show in New York and the first encounter with the info-babe show host was, "Ah drat, I thought you'd wear your jeans and straw hat."

When will the stereotype end? I'll tell you when — when farmers quit thinking the same thing of themselves. People tend to see you the way you see yourself. In my humble opinion wearing bib overalls and shredded straw hats to a farmers market is not the way to engender respect. Farmers should get a suit, take a speech class, and read some marketing and business books before heading out to prime time.

Dealing with landlords, we farmers must see ourselves as equals and demand the same from the landowner. As soon as the landowner assumes a condescending position — either covertly by body language or voice intonation — or overtly with impertinent statements, nip it in the bud. It's okay to call them out on the attitude — diplomatically of course, and with a smile. The universal assumption in our culture is that farmers are dolts and academically challenged to boot. After all, farming doesn't take brains; it just takes brawn, right? The extension of this assumption is that farming is easy. It has to be easy since only lame brains do it, right?

This is why it's imperative to deal professionally with the landlord. When you come to a meeting, dress smartly. If the landowner dresses in a coat and tie, you should match it. You will only get the respect you demand. Bring a legal pad and pen and use them — take notes. One of the things I do is send a synopsis, like minutes, of the meeting to the landlord a day after the meeting. That's a great way for the landlord to realize you're professional and on top of things. This will discourage any impetus on his part to proffer shady dealings.

It also helps to give the other party a chance to clarify. What you want is agreement, even if only tacit, that you can refer to in the future if an issue gets touchy. I always start the synopsis with, "This is my sense of what we agreed on. I very possibly missed something since we were having such a good time. If I've misperceived or missed something important, feel free to add and/or correct as you see fit. I just want both of us to be on the same page, so edit at your leisure. Let me know if this is your sense too. Thanks." A disclaimer like this creates invaluable goodwill.

Other Options

I've belabored the traditional farm lease arrangement because it's still the most common way existing farms expand or new farmers get access to land. But other arrangements worth discussing exist, as I touch on below.

One of the most underutilized opportunities is caretaking. Many elderly people who own some land need someone to come and care for their property. It may be a house and lot or a whole farm. These folks don't want to go to a nursing home, but they want someone around to mow the lawn, take out the garbage, clean the house, and maybe cook a couple of meals a week for them. A servant's heart attitude on the part of a young person can create wondrous compensation possibilities from grateful elderly folks. Sometimes this is a live-in situation or even offers another house on the property for your use.

These caretaking opportunities can go on for a long time if the elderly person or couple is relatively healthy. The key is to be patient and work with the situation you're in. We have such an entitlement mentality in modern America that too many young people aren't willing to sacrifice what's necessary to realize their dreams. But acquiring use of land in a situation where elderly folks are still on it may require a long-term attitude and vision. I'm amazed at how quickly successful business people are dismissed as having been born with a silver spoon or had some other leg up on everyone else. That their success came because they were willing to work harder, longer, and smarter than anyone else never enters the discussion. Such an attitude keeps young people from achieving what they could if they would simply put their nose to the grindstone and do whatever it takes to get to their dream. Put your opportunities into context and bloom where you can.

The change in land ownership and opportunity hasn't

occurred overnight and it won't change overnight. The way grandpa and grandma started farming won't be the way you start farming. Everything is different today. Don't curse the differences or complain about them. They are what they are.

Nobody would like to see thousand dollar per acre land more than I. Goodness, I'd buy a boatload of it. But it doesn't exist, and may never exist again. We don't get anywhere by crying about it. We get somewhere by recognizing the new realities and developing strategies to guarantee our success with the hand we've been dealt. You can't un-ring a bell. You can't roll back the clock. It is what it is, so just get on with it. If that means caretaking an elderly couple to get your foot onto a half acre of land to grow vegetables and develop the track record for a bigger deal in the future, so be it. Tears won't drop a candy-wrapped farm in your lap. Sweat, perseverance, and patience just might.

Elderly folks who grew up during tough times deeply respect hard work. They know what it is and how to recognize it in others. As a caretaker, you can do some farming on the side, even if it's a postage stamp garden. Get your feet wet. Start something. . . anything.

Another possibility is to go onto an existing operation as a manager. This is not a lease arrangement; this is more like working for the landlord. The formal arrangement can be anything from being an employee to an independent contractor to a partnership. Often a great manager will quickly gain trust and opportunity for independent non-competitive enterprises. If it's an orchard, how about some poultry or sheep under the trees? If it's a dairy, how about some pigs to eat spoiled milk? If it's a produce operation, how about some chickens or pigs to eat spoiled vegetables?

With lightweight portable infrastructure, you can add a host of enterprises to an existing mother ship. In my view, the easiest is to add poultry to a beef cattle operation. All that does is grow more grass for the cows. Talk about synergy. But don't forget other possibilities:

- A bread oven
- Woodworking
- A commercial kitchen
- Bramble fruits

- Produce
- Hoop houses/cold frames
- Bio-diesel
- Alcohol, both energy and drink
- Cut flowers
- Juice from any fruit currently grown
- Meat cutting
- A restaurant
- Farm tours
- Seminars and cooking demonstrations
- A nursery from either domestic or wild stock
- Fish
- Milling flour from grain production
- Omnivores added to any grain farm

I hope you get the picture. The list is endless, but illustrates how limitless the add-ons can be. I've just brainstormed this list in a minute, but no doubt any reader can add another ten if you spend a little time. Often, in these partnership type arrangements, the land owner already has equipment and a shop — both expensive start-up investments for newbies.

Leveraging a manager position into something bigger is doable, but may take a couple of years to establish trust and a track record. Often, older landowners who have grown lonely and tired on their places are so grateful for young enthusiasm that these relationships blossom into long-term deals. Sometimes they can even outlive the owners.

Are you in the city? Don't know where to go? How about growing plants in pots? On the roof? Stacked gardens in the postage stamp lawn? I ran into a lady recently from Edmonton who operates an urban CSA called "On Borrowed Ground." People essentially donate their back yards to her and she operates a self-sustaining produce farm on that land. Land owners have the satisfaction of a productive and beautiful backyard and she's full time farming in the middle of an urban area.

Michael Ableman, who farms on Salt Spring Island, has started growing multi-acres in Vancouver in repurposed wooden shipping bins. They're about two feet deep and four feet square.

They accumulate in the U.S. and Canada because so many metal parts are now made in China. Since our country doesn't manufacture enough to ship anything back, these wooden boxes enter the waste stream. Vancouver, of course, is a major shipping port and these boxes can be had for the taking, just to keep them out of the landfill. Michael fills them with a spent mushroom-growing medium as a soil, adds drip irrigation on top, and plants vegetables. Situated on parking lots, these box farms employ young people and grow lots of food in the shadow of stadiums and apartment complexes.

Every time I think I've seen farming practiced successfully in the most difficult place, I'll run into something even more astonishing. I've come to believe that anybody, no matter what the circumstances, can begin farming: anywhere, anytime, with any amount of capital. It just takes heart, persistence, creativity, and personal sacrifice. How badly do you want it? If you want it badly enough, you'll find a way. That's the joy of life.

Chapter 15

Germinating New Farmers

At Polyface Farm, although we're known for innovative production and landscape techniques, our most challenging and rewarding legacy may be new farmer incubation. Although we still have no business plan and no marketing or sales targets, customers continue to increase. In spite of our aversion to the notion that businesses must grow sales perpetually, or that sales growth is the only measure of success, our sales are growing, at least for now, and that creates opportunities we never envisioned.

It also creates challenges we never foresaw. The challenges, of course, are in accessing the opportunities. How do you meet an expanding market on a static production base (certain number of acres)? How do you meet an expanding market with a static work force (certain number of workers/farmers)? The answer is you don't.

When we first started the internship program about twenty years ago, we had no idea that we'd eventually use the program to vet production collaborators. We had no idea that we'd be leasing additional acreage. We had no idea we'd be selling to thousands of families, scores of restaurants, and running a herd of a thousand head of cattle or eight hundred hogs or thousands of laying hens and broilers. We had no aspirations or plans for that kind of volume.

But over the years, slowly and surely, the volume increased. More landowners in the community asked us to manage their farms. That occurred right at the time we began outgrowing our home farm — Polyface Central — which we own and which acts as the ultimate base of all our operations. Perhaps the single thing that gives me the most pleasure these days is our healing touch on more acreage. None of our goals has ever included increasing the size of our empire or production volume. What really floats our boat is watching the miraculous healing that occurs on a piece of land after we've massaged the ecological womb for a few years.

As of this writing, we lease nine properties in the area. One of them is contiguous to our home property, but the rest are scattered around the area anywhere from five minutes to twenty-five minutes away. The largest is three hundred and sixty acres and the smallest is twenty acres. This additional fourteen hundred acres of land gives us a large land base on which to operate but requires more labor to manage.

Rather than sending out labor from Polyface Central every day to move cows and poultry on these satellite operations, we realized early on the advantages of managing from on-site dwellings:

1. No transportation costs.
2. Constant oversight to see and know what's occurring.
3. Independent contractor status. Shared risk is superior to wages.
4. Personal authority and empowerment for the farm managers.

Over the years we've developed a template for formal memorandums of understanding (MOUs) to create autonomous self-employment management positions on these farms. This is a way for young people with zero equity to become full-time farmers on day one.

Most start-ups require some sort of nest egg until the embryonic farm business scales up enough to fully employ the farmer. Remember, the downside of direct and high margin marketing is that if you exceed your market by one dozen eggs or one tomato, the value drops to zero. One of the beauties of commodity farming is that the market is big enough to absorb any volume as fast as you want to expand.

Direct marketing isn't that way. Only untried people think they can grow and sell ten thousand pastured broilers the very first year. You start with a couple of hundred and then expand. But even the most talented marketer/producer won't arrive at a full-time salary within a year. The best I've seen is three years. Peter Drucker, the famous Austrian-American business management consultant, said of the business cycle that nothing becomes lucrative until it emerges from the downturn learning-curve trough in the third or fourth year.

The reason Teresa and I were able to make that precipitous, memorable, and successful, jump to full-time farming on exactly Sept. 24th, 1982 was because during our two years of outside employment and frugal living we'd accumulated a nest egg big enough to live on for at least a year. Secondly, we didn't have to pay for land — we were second generation farmers and the land was paid for. Thirdly, we grew all of our own food, cut our own fuel (firewood), and lived in our attic farmhouse apartment on three hundred dollars a month. We drove a fifty dollar car, never ate out, bought used clothes, and devoted ourselves to a dream. When the kids came along, we used cloth diapers, ground our food in a little table mill (we never bought a jar of baby food), read to them, had no TV, and encouraged them to play in the dirt.

The young people we're launching through our internship program today are direct beneficiaries of these tough and frugal early years. One of the biggest dangers in reading this book is to get overly enamored of the end result without appreciating what it takes at the beginning. I hope I'm not sugar coating the journey. It's a wonderful, rewarding, satisfying and fulfilling journey. But anything that satisfying won't be easy.

By vetting potential satellite farm managers through our internship program, we can reduce the normal setbacks associated with these kinds of high-risk collaborations. Obviously putting twenty-year-olds in charge of equipment and four hundred thousand dollars worth of pastured livestock can be disastrous if things go wrong.

By leveraging our marketing and production scale with the additional land leases, these first-time farmers can begin full-time farming immediately with a full complement of animals and infrastructure. And this is with no money down and no off-farm job supplement. That's a sweet deal. I hope that as I describe these arrangements in detail, it will awaken hope even in the most dubious hermit curmudgeon farmer's breast about the advantages of working with young would-be farmers.

For middle-aged independent-minded farmers petrified of collaborating with a young person, I have only one question: "Do you want to farm at age seventy. . . alone?" The farm that excites you at thirty will become burdensome at seventy. You won't be able to run the farm at seventy that you're running at thirty or forty. That's just a fact of life. Look in the mirror and ask yourself if you really want to farm alone at seventy years old. Now go out and embrace some young people.

Launching a full-time farming career with no nest egg and no money down is almost a fairy tale. But this is exactly the kind of opportunity we've created and the model we hope others, both young and old, will duplicate. It's a perfect catalyst to germinate fields of farmers.

For clarity and illustration, let me offer some real numbers. At one leased farm, we have three hundred head of beef cattle, two groups of fifty hogs, four thousand broilers, sixteen hundred laying hens in two eggmobile pairs, and three hundred turkeys. Plopping those numbers down on a first-season farm creates enough volume for full-time employment with no indebtedness.

This is the missing link in many farm start-up efforts. A few non-profit organizations and slow-money type investment groups support farm start-ups, and I applaud these efforts. Some buy land and lease it to aspiring young farmers. Some act as facilitators to link retiring farmers with young people. Others offer low-interest loans for beginning farmers.

But what we desperately need are successful, profitable, scalable going farm concerns ready to expand in a step big enough to create a full-time salaried enterprise at the sub-contracted independent contractor level. These fledgling enterprises unfortunately struggle, even with the outside help, because too many high-risk elements exist at once. Here's a list of the biggest risks:

1. *Lack of experience.* Just as in anything new, the learning curve takes a huge emotional and economic toll. New farmers

often make catastrophic mistakes. Without an independent nest egg or other support, novices' efforts often fail to produce enough compensating positives fast enough to overcome inexperience.

2. *Lack of support.* All of us need a network of mentors and cheerleaders. Starting a successful farm takes far more than money and time. In my own case, my parents provided much-needed early support. Teresa's family and both sets of grandparents farmed as well, which increased my support network. Failure to ensure this local support base can doom start-up farmers who have otherwise secured land and money.

3. *Lack of market.* Juggling day-to-day chores, crises, and farm development, start-up farmers often find that they can produce faster than they can sell. Add to that the reality that marketing is generally the farmer's least favorite activity and you have a perfect recipe for disaster. Suddenly the eggs pile up and other perishables rot in the field, all for the lack of a market. Again, without market assistance many of these well-meaning new farmer initiatives fail to attack the weakest link.

4. *Lack of operating capital.* The old adage "it takes money to make money" holds true in farms as well as other businesses. When mid-life folks with a nest egg ask me how much to hold back, I always tell them at least two years' worth of living expenses. Another axiom is that many otherwise successful businesses fail due to cash flow problems more than any other reason. Overcoming this cash-flow challenge is one of the big draws for community supported agriculture (CSA) schemes because the subscription sale provides operating capital to launch the season's production.

Let's take each of these deficiencies and see how our collaborative independent contractor model compares.

1. *Experience.* Although the young people we're placing in management positions on these farms lack experience, they've immersed themselves at our farm for at least a season. Arguably this is minimal, and without the next three elements even the best

short-term farming experience wouldn't be enough in and of itself to ensure success. But at least our successful interns are not being asked to re-invent the wheel. Their basic mandate is to duplicate what they've been doing for the past several months.

Because our models are far beyond the prototype stage, little innovation in technique is required. We continue to make minor adjustments, but none is truly necessary. That takes a lot of pressure off the beginning farmer. Tried and true enterprises, with proven techniques, don't require the novice farmer to possess the range of mechanical, fabrication, and husbandry knowledge necessary to launch them initially.

2. *Support.* Here's where our model really shines. At a moment's notice, one of our new collaborative farmers can call Polyface Central and ask for the cavalry. In a few minutes, a team from our core staff, along with apprentices and interns, can descend on the crisis and perform triage.

I'm not talking about trying to round up a casual friend to help get the cows in or assist in clean-up after a storm. I'm talking about the most experienced, professional group on the planet being minutes away, on call, and committed to providing assistance. Last summer we experienced the first-in-memory derecho, a windstorm with up to one hundred mile-per-hour straight-line winds. It cut a swath of destruction from Tennessee to lower Michigan, but the nexus was right here in Augusta County, Virginia.

Some people lost power for nine days. High winds blew countless barn roofs and house roofs off. In the ensuing couple of weeks, I ran some forty tanks full of gasoline through our Husqvarna 359 commercial chainsaws just getting trees off of boundary fences and trying to open access roads. We may never get all the downed trees cleaned up before they rot. On one of our leased farms, the pair of eggmobiles blew over — uphill! I kid you not. They rolled right up on their sides, axles perpendicular to the ground and chickens wondering what just happened.

The other farms had crises as well. But with the tools and experience level on our team, we went systematically from farm to farm and in a couple of days we had things functioning again. In each case, the young farmers were not only in a state of emotional shock but in a bit of a panicked paralysis as to how to proceed. Within a day Daniel or I had visited each farm, assessed the issues, created a triage priority plan, and helped our collaborators make do until we could get there to fix everything.

What's that worth? As a young aspiring farmer, to know that you're not alone, that at your fingertips you can access the best fabrication know-how, construction acumen, and innovative thinking available on the planet is simply invaluable. You don't have to feel helpless. As Daniel says, it's like having Bill Gates on your speed dial to answer computer problems . . . and he actually answers the phone. How cool is that?

3. *A built-in market.* As if the previous two assets weren't enough, our collaborators don't even have to think about marketing. They can focus their full attention on integrity production. This reduces the whip-saw feeling that plagues many new farmers. When they're in the field, they're often thinking about how to sell their production. When they're selling, they're thinking about what's not getting done out in the field.

In our model, these young people tap into the Polyface brand, or marketing umbrella. The customers and distribution already exist. What a relief. Those baskets of eggs being gathered and put away each day have a guaranteed home. Those calves out in the field have a destination. Beyond that, if one of these young people wants to add something to the Polyface portfolio, the product flows easily into the customer base. Honey? Great. Soap? Great. Duck eggs? Great.

Diversifying the production portfolio is the easiest way to leverage marketing. It's much easier to find one hundred customers who will spend a thousand dollars with you than a thousand customers who will spend a hundred dollars. The hard part is in getting the customer. Once a customer decides to invest in your venue, whether it's a website, farmers market booth, or product subscription, she wants to buy as much as she can. One-stop shopping works because it's efficient.

4. *Operating capital.* The way we've set up our collaborative relationships, these new farmers don't really need a nest egg. We

pay a per diem for moving the cow herd, and taking care of the pigs. Eggs are compensated like piece work — so much per dozen. Broilers and turkeys, due to their fast turn-around and easy cash flow, can be capitalized and paid back in a few weeks.

This is like starting a job with a steady paycheck from the first week, without having to buy the business. You don't have to buy the buildings, the land, the equipment. Everything is provided. The new farmer can begin with a positive cash flow and begin right away accumulating equity for future ownership. What an incredibly different farming start-up paradigm. It's downright revolutionary.

Contracts That Work for Everyone

Now let's go into the nuts and bolts of how these collaborative arrangements work.

The basic concept is to create customized independent compensation packages with roughly equal input between Polyface and the young farmer. Rather than hiring employees like most businesses, the idea is to let the young people bring their plan to the table and then fit it into the Polyface menu. Because the interns participate in the farm each day and are privy to our discussions, including needs, wishes, and prohibitions, the packages grow organically out of an intimate knowledge regarding the mother farm business.

Most of the progress we've made on our farm has been serendipitous. We just kind of bumble along, try new things, and eventually — often in retrospect — realize we have developed something really special.

Our first land lease occurred in spite of ourselves, not because of ourselves. A fellow in our county bought a farm. He soon realized he was over his skis, so he came to see me with an offer. He asked me to lease his farm for a year so that he could learn how to do this and then said he'd take it over again. I didn't want to complicate my life with such a proposal — besides, his farm was half an hour away — so I turned him down. Offhandedly discussing the encounter at the dinner table that evening, our first formal apprentice, an outgoing young man name Tai, latched onto the idea.

Developing a soil-building, carbon-sequestering, perennial-driven intensive grazing operation from scratch intrigued him. Tai's penchant for entrepreneurship and innovation moved him to make an offer to me — "If you'll lease it, I'll run it and we can profit share." How could I turn that down? I kind of like to do new things too, so he hooked me.

I called the new farmer back and said we'd do it. With about eighty acres of rough, poor pasture to work with, we went over and put in the water and electric fence system, built a rudimentary corral, and bought sixty stocker calves (each weighing about four hundred and fifty to five hundred pounds). True to his word, Tai ran the whole operation skillfully and faithfully. We bought the calves in late March and sold them in early November, clearing about six thousand dollars. This was about twenty years ago.

As you can imagine, the landowner and aspiring farmer, who continued working in town, didn't have time to learn what we were doing, so he asked us to come back the following year. Tai left to go back to a small twelve-acre parcel his family leased. I remember well that next summer having a phone conversation with him in which he described the scene on his pasture: ducks, chickens, sheep, goats, cows, and turkeys all in the same paddock. Even though he had to buy wrenches and pliers — the most rudimentary tools — that first season he netted a thousand dollars per acre on the twelve acres. Talk about savvy.

I'd squander this story for this book if I didn't take a moment to remind everyone reading these words that I would never have leased the first farm had I been unwilling to entertain Tai's offer. Indeed, Tai would never have been able to do it, we wouldn't have done it, and the whole land-healing extension ministry of Polyface wouldn't have started if we hadn't offered an apprenticeship in the first place.

Good things, interesting things, innovative things happen when we embrace young people and then let them gallop. They may take us places we never envisioned going, but often those places are more rewarding than the places we planned to go. Never underestimate the power of unleashed youth under the guidance of elder wisdom. That's a winning combination.

We kept leasing that farm for several years until the owner

decided to sell it and move to a foreign country. But the very week he told us the news, another neighbor — only five minutes away — approached us out of the blue with a similar offer: Come set up the system, teach me, and I'll take it over. This situation offered better land, closer, and bigger acreage. What was not to like? Empowered by the first experience and having developed a larger beef market to handle the extra production, we jumped right into the second opportunity.

Just like the first lease, the landowner never took the land back over, and eventually sold the farm in a land swap arrangement so we survived the ownership change. We're still leasing that farm and have a sweet, precious relationship with the landlords. Now we've added another eight properties, but the whole shebang developed due to the enterprising character and vision of Tai those many years ago. Thanks, Tai.

The collaborative independent farm manager agreements we use today evolved directly from that first arrangement with Tai. The arrangements today are far more sophisticated but they're still relatively simple. We call our agreements simply a Memorandum Of Understanding (MOU). It's not a legal document and generally we don't even sign it because we believe strongly that all human interactions revolve around stated expectations and trust.

When Daniel and I do farm succession seminars, one of his segments has to do with silent expectations. Those are killers in any business like they are in any family. The basic premise is that I can't be responsible for your expectations if you don't tell me what they are. Indeed, modeled after the Ten Commandments, the point is that only when I articulate, clearly, what I expect from you can I hold you accountable. I have no right to be miffed over your performance if I haven't explicitly laid out my expectations of you.

Because we're human and not God, sometimes we don't know what will miff us until we've already been miffed. If we as miffee express our miff to the miffer, hard feelings ensue, justifiably, if we never originally expressed what will miff us. I'm sure this is a grammatical abomination of the word miff, but it's such a good word here I had to enjoy stretching it. Call it poetic license.

Let's assume a guy and gal get married. He's from a family where men rule outside the home and women rule inside the home.

In his family, the men didn't do dishes. The wife grew up in a different family, where the men enjoyed kitchen duty and shared all the domestic responsibilities. After a few months of marriage, this new wife will likely get miffed. The guy won't have a clue.

"He doesn't help," she complains.

"What do you mean? I bring home a paycheck every week, dear," he responds.

Hopefully, this couple will start vocalizing their expectations sooner rather than later. I don't want this to turn into a marriage counseling book — I could write pages about personally experienced marital disharmony over silent expectations — but the idea carries over into these MOUs. Both parties to the agreement need to put down, in writing, their expectations. That protects each from misunderstandings and being miffed.

Listen to the voice of wisdom and experience. I can assure you that almost all the miffs that happen in our arrangements are due to silent expectations not being met. Here's an example of a template agreement to help put those expectations out on the table (without the most sensitive details such as dollar amounts). In italics, I will add clarification notes to keep you from being lost in the inherent pre-understanding and intimate background that precedes all of these arrangements.

POLYFACE MEMORANDUM OF UNDERSTANDING WITH SHRADER THOMAS (SWINEHERD PRODUCTIONS) FOR THE YEAR 2013: DATED JAN. 1, 2013

1. Locations:

> a. Grey Gables. *This is a 180-acre farm located 5 miles from Polyface Central.*
> b. Mitchell's. *This is a 120-acre farm located 6 miles from Polyfacre Central. These two farms are only one mile from each other, which makes them conducive to running as a single unit.*

2. Minimum expectations:

> a. Move cows when there. *We move the cows every day to a new paddock, but in addition, we move the herd from*

farm to farm on a longer rotation. This amalgamates the herd into one rather than many smaller groups. Although we have hauling costs, they are more than mitigated due to herd amalgamation. Obviously, Shrader is only responsible to move the cows if they are on one of his farms; he's not responsible to move the cows if we've moved the herd onto one of the other farms run by someone else.

b. Operate one pair of eggmobiles and move every other day (arrival early March and departure mid-November).

    1. Shrader orders feed, billed to Polyface.

    2. Polyface owns birds and pays for feed.

    3. Keep eggs refrigerated and bring to Polyface at least weekly.

c. Create and dismantle paddock fences and make sure fences are adequate to control livestock.

d. Follow and adhere to all Polyface Standard Operating Procedures. *Some of these, like the broiler SOPs noted in chapter 10, are written down. Some are not written down, but contained in the body of knowledge acquired through the intern program. This point is simply to give Polyface leverage in case a subcontractor deviates greatly from our protocols.*

e. Be aware and notify Polyface of any maintenance issue with machinery, pumps, or infrastructure.

f. Shrader will ramrod acquiring a mutt dog for the eggmobiles. *Fearing predation at these farms, we wanted to try simply using a canine presence--of any stripe--at the eggmobiles as a canine predator deterrent. The ramrod point is that Shrader takes the leadership, makes the pick, and gets the dog--i.e. he's in charge, not us.*

3. Expectations on Polyface:

    a. Deliver and vacate cows. *This means Polyface brings the cows to the farm when we think it's appropriate and we take them away when we think it's appropriate. That's not a responsibility of the subcontractor.*

    b. Assist with high risk moves.

    c. Provide a tractor, to be shared with Amanda on pigs:
        1. Routine machinery maintenance provided.
        2. You break it, you fix it.
    d. Provide 4-wheeler, same arrangement as tractor.
    e. Provide roughly five hundred laying hens for the eggmobiles. *250 apiece.*
        1. Provide one pair of eggmobiles.
        2. Provide one feed cart.
    f. Provide fencing, energizer, and water troughs for cattle.

4. Future opportunities and possibilities:
    a. Moving cows at Anne's. *This is another farm we leased, also near to where Shrader lives. We wanted him to know that if he felt underemployed and wanted more to do, he could also move the cattle if we had them at this other farm.*
    b. Pigs at Mitchell's. *The landlords on this farm have given permission to run pigs in their woods, per Polyface protocols (electric fences, paddocks, and constant moves from spot to spot). Since we didn't have the infrastructure in place to do this at the time of the MOU, we wanted to put it in so Shrader knew it was a possibility if he wanted to be in charge of those pigs.*
        1. Polyface provides tractor.
        2. Polyface provides feed buggy.
    c. Turkeys. *Since this was Shrader's first year as a subcontractor, we wanted to be sensitive to the learning curve by not piling up too many responsibilities too fast. Shrader was extremely interested in turkeys, so we put in turkeys as a possibility just to recognize that interest.*
    d. Personal complementary trials like Heritage Turkeys, just with a heads up to Polyface prior to the trial.

5. Compensation:
    a. Moving cows--$ per group per day.
    b. Eggs: $ per dozen, mediums and larges only.
    c. Pigs: $ per group per day.
    d. Shrader presents invoices to Polyface for payment; Polyface makes checks payable to Swineherd Productions.

6. Duration of this agreement:
    a. The season shall terminate Dec. 31st, 2013.
    b. Failure to complete the season shall carry a $ salvage cost payable to Polyface on day of departure.

7. Misunderstandings:
    a. Either party may call a meeting to discuss any matter at anytime.
    b. Unresolvable matters may be remanded to arbitration, but not litigation.

Since this is a first-year arrangement, it has a section called "Opportunities" that can be added in mid-season if everything is going well. Sometimes young people over-extend themselves. We've found it's better to start in gently than to jump in over your head. But by going ahead and discussing additions, a mid-season adjustment is easy.

"Anne's" is a new farm we leased shortly before this MOU was written. Often our core Polyface team runs a new leased farm for the first year to work out any bugs in water and fencing systems we've installed, and to get a feel for productivity (cow-days) and movement patterns. That initial season's shakeout creates the confidence and precedents to plan the next season with less experienced newbies. Now that we've developed so many of these properties, our expertise is much better than it was years ago. So we're apt to open it up to one of our sub-contractor collaborators mid-season rather than waiting a whole year.

By paying a certain rate per day, per group of cattle, we protect Shrader from working uncompensated days, which would be easy to do if we simply paid a salary and a warm, wet fall season extended the grass production beyond our plan. By the same token, it protects us from Shrader's inefficiencies as he learns. If we paid hourly, his inefficiency at setting up cross fences or moving water troughs could easily eat up all our profits. Realizing this is an ongoing education, the per group per diem creates an extremely nice return for an experienced operator and incentivizes novices to attain that level of expertise and efficiency.

One other thing to note in more detail is the "failure to

complete" penalty, which was in item six of the MOU concerning the duration of the agreement. Patterned after termination clauses for mobile phone contracts or penalties for early withdrawal on investment monies, we've put these amounts in for two reasons:

1. If one of these managers has a golden opportunity and wants to jump on it mid-season, he can do so by paying us the early termination amount. Some of our young farmers, for example, come from highly skilled information technology or engineering backgrounds. If mid-season they realize this farming gig isn't what they thought it would be and some lucrative offer for a six-figure position in an office somewhere lands on their table, we need them to be able to leave in good graces. No hard feelings; all is well and wonderful.

2. It establishes the fact that if someone jumps ship in mid-season, our core team here at Polyface must pick up the slack. One year, we had a family managing a leased farm tell us on a Friday that Monday would be their last day — right in the middle of the season. At the time they had thousands of broilers out in the field, two eggmobiles, a herd of two hundred and fifty head of cattle, and several hundred young turkeys. That's a lot of responsibility to abandon, without warning, to someone else. We assess these early termination amounts based on the volume and risk. Some are as low as seven thousand dollars and some are more like twelve thousand.

Here's a different type of MOU example, with Buddy and Jill Powers of Powers Farms. A couple of points to pay attention to in reading this agreement are that we had already worked with this couple for two years at another farm. This year they moved over to one of the other farms to partner with another family (Grady and Erin) who were already managing that one and living there. Hence, the shared language you'll see in the memo concerning shared equipment and carefully delineated fields where they can run their poultry operations. In other words, this arrangement indicates a second enterprise layered on a farm already layered with a separate multi-species arrangement.

POLYFACE MEMORANDUM OF UNDERSTANDING WITH
BUDDY AND JILL POWERS FOR THE YEAR 2013:
DATED MAR. 1, 2013

1. Location: Briarmoor. *This is a 240 acre farm located about 15 minutes south of Polyface Central.*

    a. Shed at Libby's — one bay for shop, one bay for storage. *Like many farms this one comes with some infrastructure we can use; in this case, a small four-bay equipment shed. We store a couple of hay wagons in two of the bays, so we needed to delineate how much of the shed Buddy could use and where his storage could be located. If you don't delineate these things, you can find all sorts of spread out uses in buildings all over the property.*

    b. Shed must be fenced out/protected from cows.

    c. Brooder may be set up right behind the shed.

    d. Broilers in hayfield.

    e. Turkeys anywhere on lower place.

2. Minimum expectations:

    a. Follow and adhere to all Polyface SOPs.

    b. Be aware and notify Polyface of any maintenance issue with machinery, pumps, or infrastructure.

    c. Broilers:

        1. Produce six thousand for Polyface and one thousand personally.

        2. Separate label. *Because the broilers and turkeys have such a quick cash turn-around, we encourage our subcontractors to own the birds. It does not require much capitalization and frees us from the tension of seeing wasted feed, for example. The separate label also enables us to extend the PL90-492 Producer Grower inspection exemption of 20,000 birds per year beyond Polyface Central. These birds carry the subcontractor's lable, but also a "Product of Polyface" green oval sticker to identify it as being under the Polyface umbrella.*

3. Further processing:

    a. Parts and pieces up to nine hundred. *Nine hundred total birds into parts and pieces--we can't take an unlimited amount, so this is part of our inventory control.*

    b. Whole bird cut-ups only as needed.

    c. Livers as needed.

d. Turkeys:

    1. Produce one batch of three hundred plus fifty personally. *Buddy and Jill have goals to eventually establish their own farm about 150 miles away from Polyface Central. These personal birds do not carry the "Product of Polyface" sticker. Buddy and Jill sell these independently. This gives them a customer base and primes the pump for their future endeavors. We do not consider this competition; we consider it complementary.*

    2. Brood four batches with broilers — one hundred and fifty per batch. *Brooding is starting the little chicks or poults (baby turkeys). A brooder is a special protective area with supplemental heat to simulate a mother hen. We brood poults with chicks in a 1:5 ratio, but since the poults grow for sixteen-eighteen weeks and are more fragile than the chicks, we only raise four batches rather than the standard eight for the broilers.*

    3. Separate label. *Same reason as broilers above.*

    4. Second batch of Polyface turkeys desired if market is available. *Obviously this decision would be made mid-season based on mortality and market. Some of these points in the agreement are simply a way to get everyone's druthers on the table, kind of like a synopsis of a meeting, or a set of minutes, in outline form.*

e. Other opportunities — ducks — Powers Farm option, proceed with on-going agreement/discussion.

f. One load gravel on main lane to hay field. *We're still talking about minimum expectations, per the above heading.*

This means that Buddy will pay for one truck load of gravel on the landlord's access lane because the poultry operation will definitely put more wear and tear on that lane. Note below that Polyface also put one load on the lane.

3. Expectation on Polyface:
   a. Share tractor with Grady. *Since one tractor is plenty for both of these operations on the same farm, we offered the tractor we'd already provided for Grady to be used additionally by Buddy. The tractor stays at the farm and the two subcontractors work out its use between them. This illustrates the idea that Polyface staff do not want to micromanage.*
      1. Routine tractor maintenance provided.
      2. You break it, you fix it.
   b. One load gravel on main lane to hayfield.

4. Compensation:
   a. Broilers:
      1. Saleable birds through Polyface are X cents apiece charge.
      2. Charge of $ per bird for processing for independently sold birds. *By renting the processing shed at Polyface Central, we fall under the PL90-492 exemption, which does not specify processing location; it only specifies production and processing authority.*
      3. Options (per pound):
         a. Whole birds: $
         b. Whole bird cut-ups: $
         c. Parts and pieces: $
   b. Turkeys:
      1. Saleable birds through Polyface paid $ per pound.
      2. Charge of $ per bird for processing for independently sold birds.
      3. Seven-week-old poults paid either way at $ per bird.

5. Duration of this agreement:
    a. The season shall terminate Dec. 31st, 2013.
    b. Failure to complete the season shall carry a $ salvage cost payable to Polyface on day of departure.
    c. Failure to complete the season by Polyface shall carry a $ severance payable to Powers Farm, due by Dec. 31, 2013.

6. Misunderstandings:
    a. Either party may call a meeting to discuss any matter at anytime.
    b. Unresolvable matters may be remanded to arbitration, but not litigation.

Ready for another MOU? This one has some interesting and unusual factors to watch for. I'll explain after viewing the contract, why we did this one. Here you go:

POLYFACE MEMORANDUM OF UNDERSTANDING WITH MATILDA MITCHELLSON (not her real name) FOR THE 2013 SEASON: JAN. 1, 2013

1. Location and Housing:
    a. Apartment at Greenmont provided ($ value). *Greenmont is another of our farms, the biggest one, located about half an hour away from Polyface Central.*

2. Food:
    a. Meals on your own.
    b. Access to staff pallet. *In the walk-in freezer, we maintain a pallet of meat and poultry designated for staff consumption. Product ends up there for many reasons. It could be a pinhole in a cryovac seal that allows a bit of frost deterioration. Generally it's a cosmetic packaging issue. Sometimes it's simply an overage issue. At any rate, anyone designated as staff may utilize the food on this pallet.*

c. Regular employee/subcontractor discount (15 percent). *This is our standard discount given to all employees and subcontractors for our regular Polyface products. Anything that any customer can purchase, these folks can purchase for a 15 percent discount.*

3. Polyface expectations:
    a. Move cows when at Greenmont (240 days plus or minus).
        1. Polyface provides fencing, energizer and battery.
        2. Move cows and handle water during haymaking and feeding.
        3. Set portable electric fencing and fix minor issues with permanent electric fencing.
        4. Non-electric fences are not Matilda's responsibility.
    b. Laying hens — six hundred in a pair of eggmobiles.
        1. Polyface provides and owns the birds.
        2. Matilda moves with Polyface tractor.
        3. Polyface provides and owns the eggmobiles and feed.
        4. Polyface provides refrigeration.
        5. Extra small eggs Polyface won't compensate.
        6. Polyface provides feed storage for one ton of feed, preferably two.
    c. Hogs at Browns. *This is a 25-acre piece of property adjoining Grey Gables (see Shrader above) divided into half-acre paddocks on which we run four batches of hogs, rotated through about ten paddocks each. Obviously in our negotiations with the subcontractors Matilda picked up this responsibility, although it could easily go to the Grey Gables subcontractor in the future. All of these responsibilities are fluid from year to year. The quickest way to squelch incentive is to make an arrangement a dead end.*
        1. Four batches ongoing is the goal, but guaranteed two batches. *Obviously Polyface must be careful about over-promising. In this step, we guarantee a conservative volume, and therefore income, but state clearly our goal is a full four-batch compliment of hogs. Expectations and guarantees are both put in*

as a guiding principle for allocating groups of hogs. This protects all parties and the intent is clear.
2. Two hundred and seventy days max (March 1st through November 30).
3. Polyface provides feed, infrastructure, feed buggy, tractor, and owns the pigs.
4. Conduct one feed conversion test at any time of your choosing.
5. Check at least every other day. *This requires a visit to see that water, feed, electric fence spark and everything about the operation is fine.*

d. Research projects *Because Matilda came from a science/engineering background, in our negotiations she mentioned wanting to do some research. Fortunately, we'd been mulling over some research needs for a couple of years and jumped at the chance to enlist her passion and prowess in this area. It was a natural fit and grew organically out of us asking: "in a perfect world, what would you really like to do?" When we ask that question, we really mean it . . . and listen.*

    1. Broilers:
        a. Polyface provides three-compartment brooder.
        b. Matilda sets up numbers and keeps data, up to two hundred and fifty-bird batches and nine field shelters.
        c. Polyface owns the birds, feed, and field shelters.
        d. Projects of interest:
            1. Feed conversion.
            2. Sprouts.
            3. Different hatcheries.
            4. Biochar/charcoal.
    2. Turkeys:
        a. Brooding and field shelter only.
        b. Try a poult-only batch.

e. Phone calls must be received and returned within twelve hours of placement.

f. Equipment: you break it, you fix it.
   1. Polyface maintains equipment for regular use.
   2. Matilda maintains infrastructure for regular use.
   3. Polyface provides:
      a. Tractor.
      b. 4-wheeler and trailer.

4. Compensation:
   a. Cattle: $ per day per group less $ for the apartment.
   b. Eggs: $ per saleable dozen, larges and mediums.
   c. Hogs: $ per batch per day.
   d. Research projects:
      1. Broilers.
         a. $ per saleable bird production post brooder.
         b. X cents per saleable bird for brooding.
         c. $ per saleable bird research additionally.
         d. BOTTOM LINE: a + b + c is $ per research bird and below, e
         e. Matilda keeps research hours to see which lower at benchmark $ per hour and Polyface pays higher of the two. *As you might imagine, this was a tricky compensation formula to work out because we haven't done this research gig yet. Since neither Matilda nor us knew how much extra time it would take to measure and record all the data for the research project, we needed a way to protect both of us from runaway costs. We came up with two billing formulas designated as points d and e. After caluclating both ways, Polyface agrees to pay the higher of the two. It sounds complicated, but it really simply articulates two ways to figure cost and designates which option will be paid. We thought it was a genius solution to a very unknown project.*

2. Turkeys:
   a. $ per poult at seven weeks (normal is $ but trial adds $).
   b. X cents per pound carcass if produce all the way to slaughter (reference: X cents for processing too).
   c. Subcontractor: all checks payable to Grazing Greenmont.

5. No independent marketing.

6. Early termination penalty: $.

7. Duration: Dec. 31, 2013.

8. Misunderstandings:
   a. Either party may call a meeting to discuss any matter at anytime.
   b. Unresolvable matters may be remanded to arbitration, but not litigation.

The interesting thing about Matilda's MOU is that she's doing a research project. That took some head scratching and pencil pushing and we'll know a lot more at the end of the season, but as you can see, by building in two compensation alternatives and delineating reasons for the choice, we protect ourselves and her. While it's impossible at the beginning of something new to anticipate every nuance, thinking through as many contingencies as possible, together, insures that all parties come to the deal with a high level of trust.

Note that none of these agreements has hourly payment rates or employee language. I've never liked wages because they don't incentivize efficiency. Job reviews then become filled with tension, with the supervisor accusing the employee of dilly-dallying. Effort assessment is inherently highly subjective, and subjective assessments lead to tension. How do you measure ninety percent effort and a hundred and ten percent effort without some sort of piecework standard, such as achieving so many widgets per hour? Many jobs aren't conducive to such objective efficiency standards, and what ensues is a subjective perception that leads inevitably to mistrust.

Note that Polyface independent contractor checks aren't paid to people; they're paid to independent business entities. This keeps us out of the tax implications and workmen's compensation nightmares that plague profitable business in America — particularly innovative small business. Please don't get me started. It's my belief that these kinds of incubator arrangements will become more and more common as the business regulatory climate hounds employers into unprofitability.

Realize that our overriding ambition in bringing on these young people is not primarily to increase our profits, sales, or empire. It is primarily to springboard self-sustaining and free-standing young farmers into their own businesses. Yes, the best ones we'd like to stay with us, on our team, forever. But trying to keep them is not our primary concern; making them successful in their own right, in their own businesses, is the over-arching goal. We're not trying to build an empire of employees or vast business structure. We want to be, as the title of the book *Small Giants* suggests, simply a small giant. By providing a platform on which symbiotic mentor/beginner relationships can flourish, we're happy to let the business morph into all sorts of different permutations.

Now here's one for Brie, our office and kitchen wizard, who's on full-time staff here at Polyface Central:

POLYFACE MEMORANDUM OF UNDERSTANDING WITH BRIE ARONSON FOR THE 2013 SEASON: JAN. 1, 2013

1. Housing: None.

2. Food: Polyface employee discount and staff food pricing.

3. Polyface expectations:
    a. Office:
        1. Handle Polyface gift shop orders.
        2. Handle Metropolitan Buying Club e-mails.
        3. Handle sales building one day per week (on Wendy's day off).
        4. Handle on-farm bulk orders (the green sheets).
        5. Book and tabulate Lunatic Tours.
    b. Polyface Farm Chef: June 1st through September 30th.
        1. Plan and cook the evening meal for everyone Monday through Friday at 6:30 p.m.
        2. Coordinate with Leanna to purchase as much as possible from Polyface gardens — keep a tally of poundage so Leanna can be compensated.
        3. Recordkeeping on expenses so we can tabulate actual costs at the end of the season.
    c. Grass Stain Tours
        1. Complete autonomy to market, schedule, and conduct anytime.
        2. Charge is $, split on commission, $ to Brie, $ to Polyface.
    d. Rabbits — help as desired. *Although Brie's MOU is heavily weighted toward chefing and office work, she like most of our staff really enjoys getting her hands dirty. We like team players who enjoy playing all the positions occasionally. Designating her first area of hard core farm work interest helps everyone appreciate personal priorities. She doesn't have to do anything with rabbits, but is welcome to inject herself into that enterprise whenever she wants to.*
    e. Garden — help as desired. *Same idea as with the rabbits.*

4. Compensation (payable monthly):
    a. Office and sales building: $
    b. Chef work: $

5. Duration of this agreement: Dec. 31st, 2013.

6. Time off: up to three seven-day mutually agreeable times per year.

7. Misunderstandings:
    a. Either party may call a meeting to discuss any matter at anytime.
    b. Unresolvable matters may be remanded to arbitration, but not litigation.

Aren't these MOUs fascinating? Note that although Brie has a delineated salary from two primary responsibilities (chef work and computer work), she also has a commission-based portion from the Grass Stains Tours. Although we haven't done it for every single staff person, we try our hardest to create some sort of personal fiefdom for each person, with concomitant commission-based remuneration. Everyone needs a fiefdom. Fiefdoms are cool, even if they're small.

    When Brie, a former intern, approached us about staying on, she knew that I'd always wanted to offer farm tours for school-aged children but never had gotten around to it because we didn't have the personnel. I see my role as simply dropping hints about what I'd like to see, offering opportunities and wish lists rather than giving orders. If an intern is listening, she'll hear a meaningful compensation opportunity. That's exactly what Brie did. She knew we wanted to offer this and stepped forward to create, market, guide, and move the enterprise forward.

    Brie loves children. She has a bachelor's degree in Elementary Education, but after a year in the classroom, she realized that the institutional setting was not her style. She came to us as an intern a couple of years after finishing college and we now refuse to let her depart . . . ever. She's shackled to the farm and we threw the key in the river last week. Ha!

    When she approached us about starting school tours, it was a no-brainer. Of course we'd be open to it. Of course we'll let you run it. Of course it can be your fiefdom. The first year she had twenty takers, the second was forty takers and who knows where it will go? She simply advertises it on our Polyface website, communicates with the teachers, schedules them, and handles the payments. We split the tour payment, but she gets the lion's share.

She's done everything from inner city school kids to cub scout troops to homeschool families. The cost is the same regardless of the number of tour participants. Each tour is a flat rate; the cost to us is Brie's time. Whether she squires around fifty or five in the tour, her time is the same. This creates an economy of scale for large groups who may not be able to afford a $10 per person charge, for example. It even rewards carpooling and busing. The rest of us don't have a clue when tours are coming. All we know is that when a bus shows up, Brie must have a tour. She leads them out through the fields, develops her own instructional two-hour dialogue, and has a remarkable rapport with teachers, parents, and students. And yes, even kindergarteners walk the mile and a half to see everything. Brie can coax anything out of anybody.

I have to tell you this story about Brie. When we do our Polyface Intensive Discovery Seminars (PIDS) in the summer, I give different staff and collaborators a chance to share with the attendees. This gives me a momentary break but also lets folks see the breadth and depth of the Polyface people and enterprises. The first time she shared, she relayed how early on in her time here she would explain to visitors, "This is how the Salatins do it."

But once she started the Grass Stains tours, she found herself saying, "This is how *we* do it." Her epiphany was that owning this business created in her a stakeholder ownership mentally with the farm. She no longer saw herself as working *for* Polyface, but as an equal owner *in* Polyface. Pardon me while I wipe these tears from my eyes. Isn't that precious? How could I be so blessed? Folks, it really is all about finding team players, releasing them to play their positions, and then incentivizing them to play extraordinarily well.

Of course, the Grass Stains tours bring additional customers, publicity, and awareness to Polyface's land-healing message. And you know what the coolest part of all this is? I don't have to do it. I can be sick, dead, or gone, and Brie still takes these youngsters, hand-in-hand, traipsing like little sponges across the lands we've healed. Hallelujah Chorus, anyone?

One final point on Brie. The tension between wages and salary is ongoing in the business world. Too often employees view a salaried position as a way for the employer to extract far more hours than fairness would dictate. On the other hand, salaries certainly incentivize efficiency as long as the employer doesn't just reward efficiency with bonuses or additional uncompensated responsibilities. Since we didn't know how much time either of Brie's primary salaried responsibilities would take, the first year she kept time sheets for each of these activities.

During her year-end review, we looked at those times to see if we were in the ballpark. Obviously if they had worked out to five dollars an hour, we'd have to revisit our arrangement. That could get tense, of course, if we felt like she was taking too long to do things on the one hand, or if she felt we were abusing her on the other. While I don't like hourly wages for anything (we have no employees working for hourly wages) they still are out there in some assumed ethereal worker never-never-land. While we don't use hourly wages per se, we're well aware that they create benchmarks, silent expectations if you will, in our work culture.

As it turned out, Brie's time sheets corresponded extremely well with our mutual ideas of a fair hourly rate, validating our original assumptions. That is not always the case, so it's best to create a safety net that catches a gross miscalculation. This safety gauge gave Brie emotional buy-in to our commitment to protect her interests. Acquiring and keeping loyal people starts with this level of respect.

Finally, here's a MOU for Leanna, who wanted to layer a horticulture operation on top of the Polyface livestock enterprises:

POLYFACE MEMORANDUM OF UNDERSTANDING WITH LEANNA HALE FOR THE 2013 SEASON: JAN. 1, 2013

1. Housing provided ($ value).
    a. Housing in Daniel and Sheri's basement.
    b. Electric and phone provided.
    c. Sheri's washer/dryer access.
2. Food:
    a. Meals on your own except for evening meals during the summer when interns are here.
    b. Access to meat and vegetables like apprentices and interns — unlimited to your garden goodies. *One of the perks of living on-farm is free access to the Polyface intern and*

apprentice food stash. This is what we call a pot sweetener because it offers gentle emotional and economic massage.

3. Polyface expectations:
    a. Oversee sanitation of kill floor and coolers.
    b. Inventory manager:
        1. Stock retail store freezers and refrigerators.
        2. Inventory freezer and give volume sheet to proper person weekly.
        3. Go to T and E Meats weekly to pick up meat and inventory.
        4. Help with bulk on farm beef and pork pick up days. *Customers who really want to save money buy quarters, halves, and wholes of beef and pork. We offer several designated pick-up days for this throughout the season. Obviously those days are extremely hectic in the sales building. Because she knows where all the boxes are, Leanna needs to be there to sort things out.*
        5. Oversee Buying Club pick and loads.
        6. Help with restaurant load up Thursdays and Fridays.
    c. Oversee brooder.
    d. Oversee Polyface gardens on commission.
        1. Royalty to Teresa and Sheri as follows (in pounds unless noted): *These amounts are shared between our households and represent the total royalty.*
            a. Potatoes   300
            b. Sweet potatoes   100
            c. Tomatoes   200
            d. Green beans   100
            e. Yellow squash   40
            f. Zucchini   60
            g. Cucumbers   100
            h. Sweet corn   200 ears
            i. Onions   60
            j. Butternut squash   40
            k. Carrots   40 feet of row for winter

2. Marketing priorities:
    a. Polyface kitchen for Brie — at wholesale. *See pricing below.*
    b. Retail store.
    c. Restaurants (through Hannah like any other grower).
    d. Buying clubs.
3. Production areas:
    a. South side of barn.
    b. Loafing area at far barn.
    c. Terrace between hoop houses.
    d. Across pond.
    e. Floating garden on pond.
    f. Main house garden.
    g. Hoop houses.
        1. 20 X 120
        2. 20 X 120
        3. 30 X 120
        4. 30 X 120
        5. 30 X 120
    h. Vineyard.
    i. Blackberries.
    j. Fruit trees.
    k. Gobbledygo spot.

4. Polyface provides:
    a. Seeds and sets as needed.
    b. Tools.
    c. Compost, mulch.
    d. Borders.
    e. Labor as needed.
    f. Irrigation as needed.

5. Compensation:
    a. Straight guaranteed salary $ payable monthly (for non-horticulture responsibilities).
    b. Horticulture compensation based on sales (straight commission).

1. Wholesale prices are as follows (by the pound):
    a. Green beans   1.35
    b. Beets   1.25
    c. Cabbage   .75
    d. Carrots   1.25
    e. Chard   2.25
    f. Cucumber   .75
    g. Leaf Lettuce   2.25
    h. Onions   1.00
    i. Sugar Snaps   2.00
    j. Green Pepper   1.50
    k. Potatoes   .70
    l. Salad Mix   3.25
    m. Baby Greens   3.25
    n. Spinach   2.25
    o. Winter Squash   1.00
    p. Summer Squash   .75
    q. Sweet Potatoes   .90
    r. Tomatoes   1.50
    s. Ear Corn   .17 per ear
    t. Herbs — at your discretion.

2. Brie buys for the kitchen as needed per agreed prices and Leanna invoices Polyface for that amount.

3. Retail store: Add up to twenty percent to agreed wholesale (per Leanna's discretion), then add twenty-five percent more for the retail markup (i.e. store price forty-five percent higher than base wholesale) *While all this may sound confusing, it's actually quite straightforward. What it does is recognize the extra care and attention needed to prepare product for retail display as opposed to picking a box of produce for Brie to use in the Polyface kitchen.*

4. Restaurants:
    a. Leanna and Hannah sell it for mutually agreed amount.

    b. Leanna gives Hannah her commission from that.

    c. Leanna gives Polyface the three percent grower handling from that same total.

    d. Leanna invoices Polyface for the rest.

   5. Buying clubs: Same rates as retail store.

  c. Workmen's Compensation provided. *Unfortunately, workmen's comp regulations do not allow us to treat the horticulture business like a separate business as long as Leanna gets any type of salary for doing something else--in this case, inventory manager. We haven't been able to figure out how to make the inventory position into an independent contractor arrangement--rules in the tax code make it difficult. Therefore, we have to treat the horticulture business, even though it is a totally independently-operated business, as part of Leanna's salary. We wish we could pay her a salary but treat the horticulture business like an independent contractor--that would far more accurately reflect the true arrangement and shared risk. But bureaucrats do not like innovative arrangements nor accounting that truly reflects accuracy.*

  d. Okay to have a horse on-site. *This is a huge concession and of course extremely customized. During our negotiations, Leanna asked: "What would you think about me having a really big pet?" We Salatins all knew exactly what she meant and acquiesced to this request, which meant far more to her than money. She keeps the horse in electric fence and moves The Colonel around the farm buildings like a big portable weed-eater.*

    1. Can't occupy grazing space needed by anything else.

    2. Polyface provides winter hay for horse.

6. Duration of this agreement: Dec. 31st, 2013.

7. Misunderstandings:

  a. Either party may call a meeting to discuss any matter at anytime.

b. Unresolvable matters may be remanded to arbitration, not litigation.

I left a little more detail in this one so you could see the level of intricacy these MOUs can delineate. While it may look tedious to lay out this much detail, remember that silent expectations are killers. Also remember that this level of delineation has developed over time as we've learned what is necessary and unnecessary.

Here again, just as with Brie, Leanna has a base salary primarily stemming from her inventory management responsibilities. But the garden enables her to create her own fiefdom. She has complete control over this portion of her income. Her first year, she gave a lot of stuff away just for the pure joy of it. She may not continue that, but it's completely up to her.

Arguably the most intricate of our MOUs, Leanna's horticulture business could morph into a full-time position if she or someone else wanted to take it there. It's her baby, and if she leaves — perish the thought — and somebody else wants to take it a different direction, that's fine.

Collaboration Celebration

How different — and liberating — this model is can hardly be exaggerated. Rather than the Salatins of Polyface being responsible for all the visioning and directioning (I know that's not a word, but it's fun to play, don't you think?), essentially creating employment slots and trying to find people to fill them, we're letting the farm business morph into simply a manifestation of the gifts and talents existing in the personnel. What a different idea for a farm and a business. Instead of the owner deciding everything and then trying to find people to do this job or that one, the actual jobs are delineated and created by the team players who come to the stadium, so to speak. It's revolutionary and absolutely liberating.

Ultimately, this is bottom up, democratized, decentralized business development. Make no mistake, if Brie left us tomorrow and nobody stepped in to take over the Grass Stains tours, we'd discontinue them. We don't feel compelled to continue any of these complementary enterprises as core business entities. They exist merely and completely at the whim of interested team players.

This frees us from feeling responsible for everything. Not only does it empower our team players, it frees us as owners from the burden of complete responsibility for what the business looks like. Who knows who will come along in the next five years? Goodness, we could have an alternative fuel guy (or gal) show up and put us into biofuels, steam engine generators and hydrogen electrolysis from pond water.

When people ask me what I want to do with this farm, I have a host of interests. But I have neither the technical skill nor time to do them all. I'm trusting that people will continue to be drawn here to create their own compensation packages with their own fiefdoms that will synergize with whatever already exists. My responsibility is to massage the team by making sure each player has an opportunity to achieve full potential.

I apologize for the length of this chapter, but it was so exciting to write I couldn't stop. I hope you found it equally exciting and inspiring. Zig Ziglar, iconic late great marketer and author of *See You At the Top*, said that "You can always get where you want to be if you make sure enough other people get where they want to be." Matches my experience for sure.

On my knees and with trembling voice I implore, plead, beg farmers to realize the unrealized potential of the people on their farms. I haven't met a farm in the world, including this one under my feet, that is anywhere close to fully developed. And I don't mean strip malls and highways. I mean fully populated with all the symbiotic resource-regenerative independent fiefdoms possible. Not one. The reason we farmers can't imagine squeezing any more from our farms is because we come in tired. When someone suggests we're not leveraging our resources we get downright defensive: "What do you mean this farm could do more? You want me to work 24 hours a day?"

But just imagine if you could indeed work 24 hours a day on your farm . . . by enlisting the bright eyed bushy-tailed self-starter energy of a young entrepreneur? That changes the whole picture. In fact, it's doable. What's holding you back?

Let's all commit ourselves to creating farmscapes peopled with integrity farmers. That'll look pretty in the fields.

Chapter 16

Food Clusters

I could call this *farming in community* but that's too trite. I use the term *food cluster* because it's more all-encompassing. The independent contractor collaborative arrangements I described in the previous chapter have gradually morphed into a food cluster — a nexus of farmers and complimentary businesses practicing a shared approach to farming values. In our case, that's integrity farming and food.

A food cluster is one of the most efficient ways to grow farmers and re-create viable farm operations. In a way, it's a localization and communitization (not a word, but you know what it means) of the full spectrum of components that makes the industrial food sector run smoothly.

A totally independent farmer selling independently to individual customers is today a freak of nature. Perhaps our Polyface farm comes as close to that idea as any, and yet I was never truly independent. First of all, I never attempted this alone. Teresa and I were married and my Dad and Mom were still quite healthy and active when we started.

All of us could drive a tractor. Dad was a genius accountant and bookkeeper and frugal to a fault. He kept a spare set of tires for the car that he would put on for the yearly automotive inspection.

Then he'd come home and put the bald tires back on. He was frugal in business; magnanimous with charity and big causes. Ditto Teresa, thank goodness. Teresa came from a family of farmers and in the early days, my father-in-law would come over with his front-end loader and help move bedding or compost.

Our neighbor Jim was also on hand to routinely provide front-end loader work. Jim also replaced a portion of the barn roof that blew off one spring when I was in college and Dad was working at his town job. Another neighbor, Oakley, often helped Teresa and me unload hay bales in the barn. In turn, I'd go help him bale and unload hay. He worked off farm, so often I'd go over and rake hay for him so it was ready to bale when he came home from his town job. When our baler broke, I borrowed his for two years until we could afford a replacement.

Another neighbor, Ray, hauled our cattle to the abattoir before we could afford a trailer. And Brown, another neighboring farmer, sold Shorthorn bulls which, in the early days, provided cheap bulls for us. We used one of his yearling bulls so we didn't have to buy one or keep it all year. He was glad to let a young bull get a bit of experience and someone else's feed while it finished growing to maturity.

Neighbors used to help neighbors like this all the time but it's seldom done any more. Brown is gone; Jim is gone; Ray is gone; my father-in-law is in his 80s and not too venturesome with his tractor any more. Oakley is ten years older than I am and needs my help more often than he used to. Generally these neighbors and fellow-farmers aren't being replaced with similar types in spirit or ability. The old community network has fractured, fragmented, disintegrated.

Also, these days equipment is far more sophisticated and expensive. That reduces sharing options. Instead of sharing, farmers rent equipment for short-term needs. The good-natured neighborly barter arrangement is almost a thing of the past. The fact that our family now rents several farms that just thirty years ago were all independent self-sustaining enterprises shows how many fewer farmers exist. In the 1960s our farm was one of a dozen within walking distance. As a child, I knew all of the neighboring farmers by name. I could ask any of them for help and they'd gladly slip over to lend a hand.

If we needed a special tool, we knew who had it in his shop and had the skills to use it. Among the neighbors, we had one guy who specialized in welding equipment. Another could fix tires. Still another had good wagons. The point is that nobody felt compelled to buy and own every item necessary to operate a functional farm. But these farms are largely gone. Their shops are gone — or else they now charge the same as the city shop for similar work. Several of these farms have been replaced with new city-folk owners who aren't even farming.

Just as farming was changing, farm production and delivery changed, too. Long ago, domestic culinary arts graced home kitchens and most girls took home economics in school; farm communities made a living selling commodities because farmers received the lion's share of the retail food dollar. Preparing, preserving, and packaging occurred in home kitchens. As homemakers abandoned domestic culinary arts for careers and recreational and entertainment pursuits, and as grocery stores replaced closely-linked farm sales, the farmers' share of the household food dollar plummeted.

Mega-industrial food processors gladly stepped in to take over traditional domestic culinary activities. The industrial economy, cranked up into high gear after two world wars, promised new wealth and liberation to people who would abandon traditional domestic drudgery. With the advent of plastic packaging, mobile refrigeration, and a wonderful new world of chemistry at their disposal, food processors offered extremely cheap (but nutritionally deficient) alternatives to the backyard garden. As these new processors and distributors jockeyed for position in price and convenience, farmers found themselves playing a smaller and smaller role in the food system. No longer could a small commodity operation survive. Lower margins required bigger volumes. Bigger volumes required consolidation. Consolidation meant less community.

How do we re-create the strength of the farm community that we've lost? I don't entirely know, but I think the Polyface collaborative network is making a good stab at it, although the farms we rent and the fiefdoms of our hub are under the bigger farm umbrella. That's certainly different than the completely independent farm businesses of yester-year, but we're arguably running these farms with more people than they had at that time.

What does this semblance of the heritage-based farm community look like in our current cluster? It starts with an anchor farm. In our area, that's us here at Polyface Central, the five hundred and fifty acres my mom and dad bought in 1961.

For years Polyface Farm has bucked every one of the trends that damaged local farming. We dared to direct market. We dared to process on farm. We dared to provide an on-farm retail interface. Over time, this created a brand name and a loyal client base. Gradually, we amassed the infrastructure necessary for a more integrated food business.

The bandsaw mill put us in the self-sufficient lumber business. Suddenly our building projects didn't depend on outsourced lumber. Our commercial chipper put us in the self-sufficient compost business. That broke our dependency on off-farm fertilizer or carbon sourcing. Direct marketing broke our dependency on the wholesale livestock auction and raw commodity brokers. Our loyal customers became our emotional and economic support network. After a while, they got so enthusiastic about Polyface that customers even started doing our marketing for us! Word-of-mouth evangelism to spread the integrity food gospel is hard to beat.

A well-equipped shop to fabricate our portable infrastructure such as eggmobiles, broiler shelters, and cattle shademobiles was certainly cheaper than a five-hundred-thousand dollar Tyson factory chicken house or Harvestore blue silo. Torch, welder, and assorted tools accumulated.

What do farmers do when they make money? They spend it. We're no different. As we pulled the plug on typical off-farm expenses, we invested our profits in better equipment and efficient infrastructure. This created an anchor of self-reliance and financial strength as we moved into the mid-90s when we began both leasing farms and running the apprentice program.

A core of lowboy flatbed trailers, post pounder, and cattle trailer, as well as new four-wheel drive tractors with front-end loaders, provided an efficient hub to run the satellite leased properties. Today, our young farmer partners on the leased farms enjoy the luxury of our top-rate equipment, shop tools, machinery, saw mill, and trailers. Any of these farmers can access state-of-the-art infrastructure and first class operators simply by calling Polyface Central.

Industrial-scale businesses amass all of their talent and infrastructure by creating enough scale to capitalize the overhead. For example, a Tyson processing plant can employ and equip a full-time machine shop because the whole facility is generating hundreds of thousands if not millions of dollars per day. By adding a network of collaborative farms, we can justify better infrastructure by spreading the overhead across multiple operations just like the big guys.

See the similarity? While economies of scale are not everything, they are something. They are real. Infrastructure overheads put more pressure on scale. Allan Nation, longtime editor of the *Stockman Grass Farmer*, says that you have to put nearly two thousand hours a year on a tractor to justify owning it. The overhead is simply too expensive to be infrequently used. The typical high capitalization cost in farming changes drastically if we can leverage those expenses across more output and income. In order to get that on a diversified, human-friendly farm, though, it takes far more warm bodies than just one farmer. You need people to need that high quality infrastructure doing thoughtful, meaningful, land-healing tasks.

That said, Polyface now owns half a dozen four-wheel drive New Holland tractors with front-end loaders. We've bought a couple of them used at ten years-old with fewer than a thousand hours on them. That's fewer than a hundred hours per year. If you depreciate a twenty-five-thousand dollar tractor over five years, that's an annual cost of five grand. If you use it only a hundred hours per year, that's a per-hour cost of fifty bucks, not counting maintenance and fuel. That's an expensive machine. Who could justify that kind of depreciation cost every time they cranked up an engine?

But if you put two thousand hours a year on that machine, its per-hour rate drops to two dollars and fifty cents. Now you can afford the machine. Unfortunately, America's current independent farmer mentality militates against this economic reality. The resultant financial difficulty — whether from debt or from lack of access to equipment — makes it much harder to launch a farm.

The small business community recognizes this principle.

Many localities now have incubators that spread office overheads across several small businesses. Numerous start-ups can share computer services, copiers, and even technical expertise to reduce office and administrative costs. It's a wonderful concept and has helped many otherwise daunting small business schemes to launch successfully.

This same concept is why I joined the Farm to Consumer Legal Defense Fund (FTCLDF), a membership organization that pools money to offer real-time legal counsel when food police harass innovative farmers and food processors. It's like a glorified insurance program. None of us individually could afford the legal expertise fighting government food extortionists requires, but by pooling our money, we create a financial kitty to dole out to whoever happens to be the prey du jour. Call this a legal advice cluster. Having an advocate at our side really helps deal with the food inquisition.

If you have to acquire all the infrastructure on day one, it cripples the business financially because from that day on, income must stay ahead of the debt load. The high initial costs create a pack of hounds that keeps nipping at your heels. Freeing a fledgling enterprise from these crippling initial costs gives some breathing space to develop experience and efficiency through innovation. This is what I call wiggle room. Every business needs it. Businesses without wiggle room soon burn out their owners.

How does this work in practice? Nathan was our first apprentice manager. He started as a Polyface apprentice and then stayed on as apprentice manager — a new position we created so we could keep him. After that second year, he did a one-year apprenticeship with Gail Hobbes-Paige, an artisanal goat cheesemaker about fifty miles away from us. As that year wound to a close, Nathan approached us about collaborating on one of our rental farms with a raw milk herd-share start-up.

Originally from Minnesota, Nathan kind of assumed he would head back there after he was finished apprenticing in Virginia to launch his small dairy and cheese operation. But as he contemplated the move, he realized the advantage of being near both Gail, his cheese mentor, and Polyface, his farm operations mentor. He also saw the advantages of having access to welders, torches, saw mills, a customer base, tractors, and land.

As a result, Nathan proposed to move our big cow herd at one of the leased farms in exchange for enough acreage to launch his little dairy. We jumped at the chance to get on-site management and worked out a land-for-management memorandum of understanding. During his second year at Polyface, Nathan had built a small one-stanchion milking trailer for a cow he acquired and milked here. He pulled this little trailer with the four-wheeler and it worked well.

With this new farm venture, Nathan purchased a hay wagon chassis at a farm auction and milled out some lumber on our saw mill to build a ten-stanchion portable open-air parlor. An enclosed box on the back held a generator and compressor for the automatic bucket milker. This portable parlor stayed out in the paddock with the cows; he moved it each day with them. Instead of the cows walking to the parlor, the parlor came to them.

It was incredibly sanitary out there on the clean pasture, fresh air, and sunshine. Nathan kept the milk in the canisters cold by pre-freezing plastic jugs of water that he placed inside. As the fresh milk hit the cold jugs, it cooled quickly inside the canisters. He started with five cows, the haywagon chassis portable parlor, a chest freezer and a couple of household refrigerators. Within one year he was up to eight cows and had secured some fifty herd shares (investor-customers for the milk).

Having seen the advantages of the eggmobiles at Polyace, Nathan bought a cheap used 3/4 ton four-wheel drive pickup with rusted out bed for a few hundred dollars. At our Polyface shop, we helped him build an eggmobile on the back and then built a trailer eggmobile to hook onto the truck hitch. This gave him capacity for about four hundred laying hens, which he procured from us at no cost since we wanted the pasture sanitation effect behind our cow herd as well.

Now Nathan had eggs to sell to his herd-share customers. Oh, did I mention that he had three pick-up points for the milk shares? One was there at the farm, one was here at Polyface, and the other one was in Charlottesville, another hotbed of Polyface customers and the nexus of Gail's cheese customers. Do you see the beauty of this market leverage? Many of Nathan's initial customers came straight out of the Polyface customer base (we publicized Nathan's offerings in our spring newsletter) and Gail's cheese clientele.

The second year Nathan expanded to nearly twenty cows with a customer (herd-share) base of more than a hundred families. He still didn't own a tractor or any equipment. He expanded into goats and pigs on some unused acreage that Polyface wasn't leasing at the rental farm. That was his own arrangement with the same landlord.

By the third year, Nathan was heading toward forty cows, still didn't own a tractor, and outgrew the acreage we were willing to trade for moving our cows. By this time, though, he had a track record and lots of connections between Polyface and Gail's farm. He arranged to lease a portion of a fairly large farm that had been a dairy a decade previously. By this time he had married Amy and they moved everything over to the new place. Just one advantage of a portable farm: when you need more land, it's easy to move your stuff and get set up quickly.

After two years at his new location, the landlord loves Nathan and has offered him several hundred acres and the keys to the kingdom. Gail has doubled her cheese facility to accommodate a new cow cheese product line using Nathan's excess milk. That gives him some flexibility on his otherwise rigid herd share marketing model. His client base is now pushing two hundred families. When he takes cheese milk to Gail, he brings back whey to feed pigs. He has his eggmobile, pigs, some lambs, goats, and one of the prettiest little Jersey dairy herds anywhere.

In the latest development, another former apprentice, Jordan from New York, is coming back to Virginia to join Nathan in a partnership. Jordan has gradually been moving forward with pastured poultry in his own farming endeavors but sees the advantage of joining forces with Nathan. And he has a tractor. Finally, after about five or six years of farming, it looks like Nathan might get a tractor. I'm belaboring that because in addition to being funny, it shows how collaboration can enable someone to start on a shoestring.

When Nathan was at our rental farm, he needed a tractor from time to time. We'd take ours over for him to use. During the winter, it was there anyway in order to pick up big square hay bales to feed our cows. It's not like he never used a tractor; he just didn't need to own one. That's an important distinction.

By joining forces, these two former apprentices will make one and one equal three! Nothing could make me more proud than to see this kind of collaborative synergy. I wish them every success and am totally confident they'll make a fine profit farming.

For those who see the glass half empty, let me scare you to death. Nathan and Jordan are located only fifty miles from Polyface Central and marketing into one of our hottest client bases--Charlottesville. What if they take some customers away from us? No doubt, they already have. Do I fear that competition? No. I'm thrilled that folks in Charlottesville have more options and more access to land healing, life giving food. Let's say these two go-getters take away $100,000 in business from Polyface. If that happens, it will just verify the power of this model and more people will want to pay me to tell it. I'm happy to let the farm business morph into more education, more tours, less production, different production--what a small world if it can't accommodate everyone who wants to heal the landscape and produce integrity food. They're not a threat. They're simply creating more buzz.

How about another practical story? Let me introduce Ben from Maryland. After his year of apprenticeship, we leased a new farm and didn't have a manager for it. We hired Ben, on salary, to mentor a new couple at that farm and spearhead the fencing and water development. Of course, he used our post pounder and tractor throughout the process and we helped in the layout. He trained the new couple and shepherded them into the process.

The next year, one of our interns decided to come back and be a Polyface subcontractor. He had a nest egg and wanted to actually buy a small farm, which he did. He hired Ben to come and help him get the infrastructure developed. That became Ben's third year working within the Polyface cluster. With that year completed, Ben looked at our operation and made us an offer to grow Freedom Ranger broilers for us and farrow hogs if we could negotiate nearby land for him to use. (As a side note, Kristen, from Virginia, had become an intern and had then caught Ben's eye — just giving you some foreshadowing here.)

We went to the neighbor and negotiated a no-rent arrangement for Ben to grow the chickens and farrow the pigs. The Freedom Rangers are the French Labelle Rouge (Red Label) genetics for broilers that grow about twenty percent slower than the American industrial Cornish Cross. We had grown a batch one year to see how they did and offered them as a choice to our more discriminating customers.

Ben saw an opportunity to build his own subcontractor business growing these birds and farrowing pigs (at the time, we were desperately short of piggies for our grow-out operation). The neighbor said yes and Ben was up and running. Concerned that these two fledgling enterprises wouldn't provide a full-time salary the first year, we proposed a trouble-shoot retainer for Ben for all the Polyface operations: We'd pay him a set amount, which essentially funded his capitalization for the farrowing and chicken operations, if he would be on-call, like an emergency intervention technician.

Since he'd been working closely with us for three years, we knew Ben could figure things out and help us deal with panic calls like, "The cows are out!" or "The water isn't coming into the trough!" He lived here at Polyface Central in the bunkhouse with the interns, which was demeaning for him now that he had left the program, but he got room and board along with close scrutiny of Kristen. I did mention her, didn't I?

Kristen stayed on as our first product inventory manager. Ben invested his savings in a bandsaw mill just like ours. He very much enjoyed running the mill and saw it as a nice income safety valve if he could do custom milling for people in the area. The mill is portable and he can tow it to the job site instead of the logs having to be delivered to the mill.

Things progressed with the neighbor and Ben negotiated, on his own, an arrangement at the neighbor's house. The land Ben used to run his chickens and hogs is separate from where the owner lives. Today, Ben and Kristen have celebrated their first year of matrimonial bliss, living at a vacated house on the neighbor's place (just a mile away from Polyface as the crow flies) in exchange for helping the landlord a few hours a week with his farm work. They've abandoned the farrowing enterprise but created a thriving raw milk herd share business with eight Jersey cows. And yes, they took a few customers away from Nathan, but in the aggregate, added many, many more to the whole "raw milk revolution" (title of a book by David Gumpert).

Ben and Kristen have continued raising the Freedom Rangers, which they process themselves and sell through us. A few they sell independently. Opting for door-to-door milk delivery, half of their clients are Polyface customers and half aren't. In the three elapsed years since Nathan started, a whole new group of raw milk-desiring Polyface customers have entered our food tent. This year, noting the delivery overhead, they began offering their clientele the full Polyface inventory and have added that option to their door-to-door delivery. This year Ben's brother is coming to apprentice with them in hopes of duplicating their success.

And Ben's saw mill? It's been busy all spring generating a wonderful financial cushion. "Money's sure not a problem," he gushed when he popped in a couple of days ago. Music to my ears.

I could explain some of the bad decisions both Nathan and Ben have made. I could even pull a couple of "I told you so's" — but why? The big picture is that they incubated their fledging enterprises in a supportive nest that enabled them to make mistakes without sinking. That's what this is all about. They built their confidence, their relationships, and their track record through attention to detail, faithfulness, and perseverance.

What's even more wonderful than telling the Nathan and Amy and Ben and Kristen stories is that we have several equally impressive former interns and apprentices who have gone on, even on their own, to successful farming operations. You can meet many of them on our website at *www.polyfacefarms.com*. I believe that all of them, in their various stages of success and terror today, would agree that the launches I've described here have enabled them to go further faster. I couldn't be more proud of how these numerous young people have gone on to bloom in their various places. If I told all their stories, it would make this far longer than it should be. Maybe their stories could make a separate book sometime. Sounds like a good idea.

Complimentary Food Cluster Businesses

Thirty years ago when people asked me to describe my vision for Virginia's future food and farming systems, I answered, "Thousands of independent, autonomous neighborhood farms like Polyface selling directly to their communities."

I no longer think that's realistic or even appropriate. Why? In short, because very few farmers want to fool with marketing, processing, and distribution. I certainly still have a vision for thousands of farms like Polyface, but not to the self-standing, independent extent I once imagined. I see a far more collaborative arrangement. Perhaps we could call it mutual interdependence. With the advent of large-scale industrial centralized food and farming systems, the three food system essentials of marketing, processing, and distribution, on a community scale, have largely disappeared. More often than not farmers can't get Genetically Modified Organism (GMO)-free chicken feed. Sometimes a feed store doesn't even exist within a hundred miles.

Abattoirs (slaughterhouses) that used to dot America's landscape, an integral part of nearly every community, no longer exist. Farmers routinely drive a hundred miles or more, one way, just to get a hog butchered. This adds an exorbitant cost to local food, making it look elitist and unrealistic for the masses. Actually, it's an artificial civilizational abnormality created by cheap energy, prejudicial food safety laws, consumer apathy, and government market intervention.

Faced with our abattoir possibly closing several years ago, Teresa and I found a partner to purchase it, holding onto this vital link in the food-to-plate chain. That saved eighteen jobs, a local business, and the opportunity to direct market for all the farmers in this part of Virginia. Fortunately, the food cluster we've developed created enough economic leverage for us to be able to invest several hundred thousand dollars in this small abattoir and keep the doors open.

True and Essential Meats in Harrisonburg, under the management of my partner Joe Cloud, custom processes beef, pork, goat, and lamb for scores of farmers in the area. In fact, in 2012, Joe's wife Brydie shepherded a collaboration with Virginia officials to launch the Small Processor Apprenticeship Program, the first of its kind in the entire nation. Our first apprentice started in the fall of 2012 and has completely energized the older employees who often feel like a dying breed. The energy and vitality that young people bring to a business cannot be overestimated.

When GMOs first became an issue, the local feed mill mixing and grinding our poultry feed refused to carry GMO-free feed, arguing with us that it didn't matter and there was no difference. We found a small non-functioning mill on a Mennonite farm in Stuarts Draft and asked them if they'd be willing to do GMO-free poultry feed for us. It turned out that the son, who was running the farm, considered the mill an albatross. His father, who had operated the mill for chickens prior to the big vertical integration in the poultry industry, hounded the son to keep the mill operational and the building painted even though it was not being used.

With our offer, the mill came out of mothballs and the ecstatic son turned some income on what had become a farm liability. In short order, the reborn mill, named Sunrise, generated a substantial amount of the farm's income. That GMO-free market in turn pushed nearby farmers to switch from GMO grain to GMO-free grains, impacting hundreds, and now thousands, of acres and putting more money in the farmers' pockets, which in turn added revenue to the local farm economy.

A few years later, another family took over Sunrise as a full-time job for one of their sons. By this time, we became so busy that it became difficult for us to drive over to Sunrise to pick up feed. I asked the owner if he would consider buying a bulk delivery truck. His eyes sparkled and he agreed that, "It would revolutionize the business. But I'm leveraged so far on this mill I can't afford anything more."

"What if I offered you a no-interest loan, payable in two years, so you could buy the truck?" I queried.

"Oh, that would be a real blessing," he responded. Before the week was out, he located a truck in North Carolina. He took my check, went to North Carolina, and drove the truck home. And it did revolutionize their business. It extended their reach both in volume and distance. In fact, because the mill became a going concern, when Countryside Natural Products decided offering organic feeds would be a good strategic move, the existence of Sunrise actually made it doable. Certified organic, Sunrise quickly became the GMO-free go-to place for the entire mid-Atlantic region.

That no-interest truck loan was probably one of the best investments I ever made. Daniel had been making most of the Polyface feed runs before, and he was gone a day and a half each week getting feed. With the advent of the truck, he suddenly added seventy-five days a year to his farm work availability on-site at Polyface. Wooo-hooo! That freed up Dad (me), which made life much better.

Today, Sunrise is certainly one of the larger GMO-free mills in the nation, employing several staffers to serve an area extending a couple hundred miles out. In turn, the clients who patronize Polyface and all the other farms who use Sunrise are directly touching thousands of acres to keep them GMO-free.

Each time we add another farmer to our cluster, this sphere of influence grows. We can heal more land, put more money in farmers' pockets, and sustain community-based farm infrastructure more effectively.

We collaborate with half a dozen other farmers in the area that offer products complementary to ours. For example, Golden Acres Orchard near Front Royal is one of the oldest biological apple orchards in the country. The founder, A.P. Thomson, was featured in Mother Earth News magazine as a Plowboy interview in the late 1970s. Well spoken, entrepreneurial, and innovative, A.P. certainly mentored me in the ways of biological farming.

We began drinking his flash pasteurized apple juice. It's the real deal. An inch of sediment on the bottom of the jug lets you know it's the real deal. In fact, you have to pace yourself when you chug-a-lug two glasses in January and suddenly realize, "Oh crud. I just ate six apples." Then you spend the next day working through the indulgence event, if you know what I mean.

Anyway, A.P. eventually passed away and his son John took over the orchard. Like many of these successional events, in his waning years A.P. hadn't kept the orchard up to its full potential. When he took over, John ripped out decrepit trees and rejuvenated the orchard, but in doing so, neglected some of the marketing. We saw the sales struggle and offered to add their juice to our product queue. John was overjoyed. Today, Polyface moves some seventy percent of the Golden Acres apple juice in a true win-win business relationship. Our customers get the advantage of this extraordinary product, John doesn't have to market it, and gets to spend more time on what he loves: working with the trees.

Since our delivery truck goes within a couple of miles of the orchard on our northern Virginia runs, Richard (our driver extraordinaire) simply swings by on his way home and picks more up from time to time. This efficient delivery keeps shipping costs and carbon footprints down.

Just up the road from us, Mountain View Dairy Products makes artisanal cheese. Their products go on our delivery truck to the restaurants each week. They were already marketing to several of the restaurants we serviced and it didn't make sense for both of us to visit the same client. They began putting their products on our truck and it was better for everyone. We have not yet added them to our buying club queue. Several nearby vegetable growers use our local restaurant marketer, Hannah, to sell their produce and berries. Hannah works on commission for us and for this group of growers. She's making the calls anyway; why not leverage the time and person with additional offerings? The chefs like the one-call cornucopia.

This arrangement enables the chefs to order beef, pork, poultry, eggs, chicken, turkey, cheese, vegetables, fruits, juice, and mushrooms at one time, from one interruption. On a micro scale, we can rival the variety and efficiency of a mega-giant like Sysco. Isn't that exciting?

The juice, cheese, vegetables, and berries go on our delivery truck, saving all these small growers duplicate delivery runs. We're all selling to the same places anyway; why not consolidate and put more pounds on the delivery vehicle? The industry employs countless experts to figure out these efficiencies. Those of us in the local food business need to appreciate where these leverages can occur to create our own economies of scale so we can better compete with the industrial sector.

Beyond The Common Model

At the risk of offending farmers' market devotees and Community Supported Agriculture (CSA) diehards, I suggest that the food cluster concept launches more local food faster than anything I've seen. The CSA is cumbersome because the client has to buy a pre-packed box. The lack of choice relegates this marketing scheme to a very small echelon of potential buyers.

The farmers' market has its own weaknesses that keep many farmers out:

1. Market rules, which often stem from cronyism and insider shenanigans.
2. Market commissions, which can be substantial.
3. Inconvenient times; Saturday and Sunday are days for people to stay inside and pad around in pajamas, not to brave traffic and parking to go shopping, which inherently reduces the customer base.
4. Health department and food police exposure, leading to snooping, poking, harassing.
5. It's often more social than serious buying.
6. Speculative risks. Vendors take things and hope they'll sell. They then need a second market for the unsold goods.
7. Farmers must leave their farms to man booths.

My benchmark for successful farmers' marketing is that if you're not selling at least two thousand dollars in product per event, it's not worth your time. Here's my question: if you spent fifty days a year either creating a food cluster or on hard core marketing — knocking on doors, chasing down leads — would you be better off than you will be with your year's farmers' market sales?

Think about fifty full days a year, leaving the farm, concentrating on marketing. I don't know very many farmers who are investing that kind of time in marketing. My sense is that an honest answer to that question would turn most farmers' market vendors into food cluster advocates.

As the food cluster develops, it can add shared accounting services, access to government paperwork expertise, and office equipment collaboration. Collaborative production and marketing are only the beginning of the opportunity. Permutations on this theme pop up in amazing places.

Have you heard of electronic aggregation? I'm convinced that the future of the local food system is going to be in duplicating locally what Amazon.com has done globally. By that I mean using Internet shopping cart type software to eliminate brick, mortar, and cashier retail interfaces. Maintaining a physical shopper interface is expensive. This includes the time at farmers' markets.

From Craig's List to eBay to Amazon.com, electronic inventory aggregation and marketing are here to stay. At Polyface, Sheri has developed customized programming for our Metropolitan Buying Clubs (MBCs). Using online shopping cart design, we can offer thousands of families, with the push of a button, our entire inventory in real time. Our clients can shop at their leisure, cherry pick from our entire inventory, and see it arrive in their locale at a pre-arranged time.

Internet sales eliminate speculative marketing at the retail interface. I love it when we localizers co-opt something developed for globalization. Variations on this theme now include virtual farmers markets. We participate in two of these. One is Relay Foods centered in Charlottesville, which sources from dozens of farmers within our food-shed, offering an electronic local supermarket. No warehouse; no bricks; no mortar; no cashier. Streamlined and chic.

Mark Lilly has become a bit of a regional icon in Virginia with the Farm-to-Family idea. He bought an old school bus, stripped out the seats and re-fashioned it with rustic decor, using burlap, chicken wire, and rough-sawn lumber. His goal was to approximate a portable old-fashioned country general store motif, complete with a pot-bellied stove and two rocking chairs in the back. He uses the Internet to publish his stops and schedule through Richmond so people can come out of the big office buildings and shop during their break. He sources from a wide variety of foodshed farmers: dry goods, vegetables, fruit, baked items, canned good, condiments, eggs, poultry, and meat.

No question, a person could fairly easily live off the food offered from either of these retail interfaces. People routinely ask me why Polyface doesn't push to get into conventional supermarkets.

Folks, we don't need supermarkets. They've only been around since 1946. We don't need them. Treat them like a bad habit. We do collaborate with some natural foods retail stores, and we don't think they're doing the wrong thing. But their high brick-mortar-and- cashier overheads create sticker shock at the register.

If we leverage the food cluster concept with electronic interfacing, we streamline and economize the local food model so it can compete head-to-head with a cumbersome, nutrient deficient, opaque, ignorance-dependent industrial supermarket system.

Oh, transparency, you ask? How do you get that on the Internet? Easily. With Relay Foods we organized an info-tainment day. They chartered a bus and offered a day with sack lunch at Polyface to their clients. I led these folks on a two-hour hay ride around the farm so they could meet the farmer and see the production systems. They enjoyed eating good food, making new friends, and experiencing a pleasant day in the country.

The whole idea of a food cluster is to accomplish through sharing and leveraging relationships and expertise what the industrial food complex has accomplished through volume. If energy costs escalate, soil depletion continues, and water becomes more precious, the local food cluster concept will spin circles around the mono-speciated segregated externalized-cost irradiated deficient industrial system. As far as I'm concerned, the conversion can't happen fast enough.

Here's to a food cluster future. Bring it on.

Chapter 17

Stacking

I've purposely shied away from using the term succession for this chapter because too often that term is pigeon-holed into estate planning. Whenever you attend a succession seminar, it's heavy on legal instruments to actually transfer a farm to its heirs. While that's certainly an important topic and greatly complicated by evil inheritance tax law, my experience is that too often the actual legal transfer isn't the biggest problem.

The big problem is that the heirs don't want to farm. And perhaps an even bigger problem is that the older generation didn't create a farm business model conducive to opportunities for the younger generation. This book has a far bigger theme than running internship programs. I purposely decided to marry succession and internship because too often the next generation doesn't want the farm and too often the farm has already been transferred to non-farming owners, either inside or outside the family. Scarcely a day goes by that some 40ish or 50ish person doesn't ask me if I know a young person who could come and do something with their family farm. Invariably, these middle-aged folks grew up on the farm but left early for work outside of farming. Now they desperately need someone to hold the land together.

Often these folks, who grew up thinking Dad was

under-actualized, have a sudden rude awakening when they realize that the barn roof doesn't stay on by itself, the fences don't remain upright without intensive maintenance, and the fields grow up in brush and weeds if nobody is around to keep them productive. They discover that the reason it's a farm and not a wilderness area is because a farmer has done something to — and with — the land.

So what happens if you know nothing about farming but you end up with a farm? Well, the first thing to do is to think creatively about what you might do to leverage that land in different ways. I'm going to lay down some principles that create room for additional salaries from a given land base. Certainly these principles could be applied to other businesses, but I'm going to use the farm as my template.

Most farms today are single-commodity production units. At the most, they grow only a couple of things and rotations are simple. Perhaps they grow corn and beans or oats and rye. Maybe it's a livestock operation driven by raising cows and some hay. Or a fruit tree orchard with cherries and plums. You get the picture. In a low-margin commodity program, scaling up to create additional salaries appears to require additional land.

As a result, farm children who grow up yearning to join the operation develop a covetous spirit toward the neighbor's land. This isn't only a spiritual problem, but it also divides the community. It's just not neighborly to covet your neighbor's land. The problem is in the assumption that the only way to expand is to add more of the same simple resources — more land — or commodities — more of whatever the farm is currently producing.

In my book *Family Friendly Farming* I go into this in far more detail, but for this abbreviated discussion let me point out the sequence in the typical farm family. Dad and Mom are in their late 30s or early 40s when their teenage kids begin showing an affinity for possible life vocations. Let's say one child shows interest and ability to farm.

But Dad and Mom politely point out that the farm won't support two salaries. We'll call the aspiring farmer son Jim for this discussion. Jim now has a decision to make. Being discouraged about his prospects for making a living at home, the farm door closes. He heads off to college and goes to work for a big company.

He marries Jill and they settle down in their urban or suburban nest. Their two kids join Little League and Jim joins the Rotary Club. Mom and Dad are fifty-five and Jim is thirty. Life is good.

Fast forward ten years. Mom and Dad are now sixty-five and Jim is forty. He's got a 401K gaining momentum, up to four weeks of paid vacation, and some seniority in the company. The kids enjoy pizza delivery during the Super Bowl and have some soccer trophies stashed on the living room mantle. They've managed to squeeze Soap Box Derby into their activities, too. After Thanksgiving dinner, Dad (now Grandpa) and Jim take a walk out to the barn and have "the *farm* talk."

Silver-haired Dad rests his hand on Jim's shoulder, looking imploringly into his son's eyes. "Jim, I know how you loved this farm. Well I'm sixty-five now and I feel myself slowing down. I don't think I can keep the farm up anymore. I'm ready to cut back. Your mother and I are financially okay with retirement — we don't need much. But we miss the grandkids. Would you come back home? We'll work it out so you can have the income."

Jim is surprised by the unexpected request, hemming and hawing until he confesses, "Well, Dad, you know I'd love to, but the income wouldn't be nearly as much as I'm making now and I'd hate to rip the kids out of school and their sports activities. With the kids entering their teens Jill has new energy and freedom and wants to travel more. I just don't think I could do it to my family."

That conversation happens all over the world countless times a year. As farmers age, it's happening more and more. Of course, you know how the story ends. Fast forward ten more years to when Dad and Mom are seventy-five and they physically can't do the farm anymore. They lease it out to a neighbor or some big land management company. Within the next ten years, Mom and Dad will pass away, leaving the farm in an undivided interest to Jim and his sister. Now what? They're now sixty years-old themselves. Their kids are thirty and completely urbanized, playing Farmville on Facebook maybe, but not thinking about *real* farming.

This scenario plays out in thousands of different permutations, but it's always the same. The land ping-pongs around as executors, renters, potential buyers, and managers vie for access and advantage, and not always in a productive way.

The whole sad state of affairs developed because the farm never exhibited a model conducive to creating additional salaries from the same land base. Herein lies the great opportunity, I think, to break this all-too-routine cycle and emerge into something more vibrant and prosperous for all concerned. While it's true that land ownership needs a degree of fluidity to create openings for new people, I want us to think not in terms of ownership, but in terms of creating additional salaried opportunities on a given land base, regardless of its ownership.

As I've mentioned, European tenure, which royalty has used for centuries, bestows ninety-nine year leases on families. That model actually has more continuity than most land ownership in the U.S. Sometimes land ownership can be a fetish that keeps us from grasping the bigger need, which is land management continuity. A corollary is the cultural asset developed when knowledge of a place and technical knowledge of a vocation passes on to the next generation.

Imagine if two-thirds of the electricians in the U.S. went out of business within thirty years and only one in ten had a replacement with a skill level comparable to the retiring electrician. That would create a real hiccup in electrical service, don't you think? That's exactly what we're facing in farming. We're losing farm expertise by the bucketfuls. Agrarian knowledge is draining from our culture, impoverishing us for who knows how long into the future. Only time and experience can rectify this loss.

When government policies and regulations make it extremely difficult for master electricians to hire understudies, not only does it impair my chances of hiring competent electrical technicians, but it also diminishes society's pool of knowledge. Certainly some things go into obsolescence. We don't need as many people to know how to make buggy whips as we once did.

Many people would argue that we don't need as many people to know how to farm anymore either. After all, the thinking goes, since industrial farming has become so efficient, we don't need very many small scale farmers these days. But I maintain that the food produced from the industrial paradigm isn't fit to eat. And the process definitely doesn't grow soil or earthworms.

Wendell Berry captures the essence of land stewardship well when he says that in order for good farming to occur, it requires knowledge of the land. For the land to be loved, it needs lovers that know its interests and needs, and where and how to caress it, if I may add to the metaphor. That takes intimate knowledge, acquired over time. And that is precisely the nature of multi-generational information transfer. I rest my case. Berry presses his case by saying that in order to be well farmed, the land must be well known. A person, he argues, can only know a limited amount of land well. This principle may explain the difficulties of polygamy. Ha!

All of this brings us to this fundamental question of which farm model promotes more incomes from a given land base. How does a farm generate more salaries without gobbling up the adjacent farms? I don't have all the answers, but I'd like to put some ideas on the table from my own experience.

Today's farms need to be synergistic business incubators. We need to be able to add enterprises without diminishing the mother ship — the main economic activity. Rather than expanding the existing primary enterprise horizontally, it's best — most sustainable, satisfying, and lucrative — to either diversify vertically or add complementary enterprises.

D. Howard Doane, an appointee to President Herbert Hoover's agriculture task force, wrote the book *Vertical Farm Diversification* in the 1950s as a model to stimulate farm income without adding land or equipment. For example, if you're producing apples, you could begin also turning them into apple cider vinegar. Similarly, you could grow your own replacement trees. Another option is to compost the pulp from the vinegar and use it for the fertilizer, so now you grow your own fertilizer. Doane described diversifying above the point of production (like the vinegar) and below the point of production (like the saplings).

Obviously if you're producing eggs, hatching your own replacement pullets would be diversifying below and making frozen quiche to sell would be diversifying above. But too many farmers don't think like this. They're frozen in their commodity, whether it's wheat, beef, or apples. Diversifying like this naturally seems frightening at first because it often involves learning a whole new skill set. Growing apples is far different than fermenting the juice.

They truly are separate enterprises and most of us can hardly run one or two successfully, let alone several.

That, dear friend, is exactly the point. Diversifying vertically to the input side or the output side, commonly called value adding, *is* a whole new business. And *that's* a perfect opportunity for the next generation. Throughout this discussion, the principle of adding enterprises doesn't have to mean that the current farmer has to do all the new enterprises. But the point *is* to make more jobs, not fewer. The goal is to create more opportunity, to have a sense of abundance rather than scarcity, as business and leadership expert Stephen Covey points out in his influential book *The 7 Habits of Highly Effective People*.

Often the next generation can create a whole new salary simply by diversifying within the existing main enterprise. Adding a composting operation for fertility is one of my favorites. By eliminating the fertilizer bill a young person can create her own salary, and maybe more if the surplus is available for outside customers, too. Hatcheries, seed saving, and nurseries are all similar opportunities.

If you're growing corn, how about making some corn meal? Cornbread? Pre-packaged cornbread mix? You can take this as far as you want. What about starting an on-farm restaurant? As a farmer, you may loathe the idea of a restaurant. But that might be just the ticket into the farm for a son or daughter, or young friend. Maybe even an old friend.

Beyond vertically diversifying, adding another complementary product is another way to create more income from the land base. Of course, my favorite for this is the pastured broiler model. Any operation growing herbivores is a natural fit for pastured broilers. The cattle, sheep, or alpaca operation (let's be herbivore-inclusive) needs pasture. Let's assume that's the mother ship.

A pastured broiler operation can be layered onto that pasture without diminishing any income to the herbivore operation. In fact, it's synergistic because the poultry manure helps the grass to grow better. Any grass-fed beef cattle farmer should welcome a pastured poultry operation — no rent necessary — due to the fertility stimulation. Suddenly the land, which represents the farm's greatest equity, is producing two things instead of one.

Pastured broilers can be layered under orchards, vineyards, grain farms and even vegetable operations. Obviously these kinds of enterprises work best if the farm practices direct marketing. When Daniel began his rabbit project at eight years-old, we were already direct marketing beef and poultry, so it was easy to add the rabbits. Ditto for Rachel's pound cakes and zucchini bread. We were already making deliveries, so it was simple to add another box containing her baked goods.

Direct marketing opens up this synergistic enterprise door like nothing else, partly because it allows embryonic development. Most commodity markets require huge capital outlays to produce the volume attractive to a buyer. Additional enterprises need to be low capital start-ups so they don't threaten the mother ship. We've all heard the axiom that "the middleman makes all the profits." Rather than demonizing those profit-sucking middlemen, how about joining them?

If that's where the profits are, I want to be my own middleman. Those notorious middlemen wear many hats — processors, distributors, marketers. Each of these hats has subsets, from graphic artists to bookkeepers. Why not employ all of these people on the farm, rather than the farm being a low margin supplier for all these external enterprises, often housed in the city? If we're trying to create opportunities on the farm, sometimes it's more efficient to take an existing product directly to market than it is to grow more products.

For example, the going commodity price for finished hogs might be sixty cents a pound, offering a farmer a five dollar gross margin per hog. At the meat counter, however, that hog is worth a dollar forty per pound. If the hog is pastured, it might be more like two bucks a pound. By capturing those additional dollars through marketing, processing, or distributing — or all three for that matter — the farm and farmer can greatly increase our gross margin to two hundred dollars per hog — or more. Suddenly we don't have to be such slaves to production.

Looked at another way, imagine a four-legged stool. While the ratio differs dramatically from commodity to commodity, for sake of discussion, let's assume that each of these legs garners a quarter of the retail dollar. We'll name these legs production, processing, marketing, and distribution. Most farmers only have one leg, which creates an unstable stool. The production leg is subject to the vagaries all farmers sit at the local diner and complain about: weather, price, pestilence, and disease.

At most farm conferences, all of the workshops will be about these production challenges. Tragically, farmers can do precious little about most of them. On the other hand, if you're processing, a drought doesn't destroy your stainless steel tables or walk-in coolers. Grasshoppers don't eat up your delivery truck. Retail prices don't fluctuate very much, even when commodity prices swing by a hundred percent. All of these assets can be used in some capacity to fill the gap when challenges strain normal operations. By creating a four-legged stool, a farm insulates itself from the vagaries that plague the business.

One of the reasons raw commodity prices fluctuate more than final retail is because farmers are most able to defray actual costs to another day. The external marketing, processing and distribution elements can't defer labor costs, infrastructure maintenance costs, or depreciation. The phone bill has to be paid when it's due. Office staff will not abide a faulty toilet. Industry businesses can't run bald tires on their delivery trucks. Regular businesses can't skip a few paychecks. Farmers, on the other hand, can and often do mask real expenses.

The farmer can defer the cost of maintaining soil fertility, or for that matter, soil. He can take a job in town to compensate for low or no pay checks. He can squeak another year out of the fence or tractor. He can put off painting the barn. The retailer can't put off painting the store front, replacing depreciating refrigerators, or fixing the leaky roof. The middlemen, then, push all their bargaining clout down to the farmer who absorbs the full brunt, takes what he can get, and hopes next year will be better. It's sad, really. This is why I want farmers, through direct marketing, to be price makers instead of price takers.

While it's true that direct marketing becomes more difficult in lightly populated areas, most farms aren't located in those areas. I wish I had a strategy for every single farm or ranch, but I don't. If you live a hundred miles from a Coke machine, direct marketing can be problematic. But rather than being paralyzed because we can't fix every problem, let's focus on the ones we can fix. Just because everyone doesn't enjoy the same proximity to markets doesn't mean nobody should access this opportunity.

I look at it this way: if every farmer who could adapt stacking enterprises would adopt them, the whole farm-and-food picture would change so fundamentally that we can't even imagine the opportunities that would be created under a new reality. If everyone within half a day's drive of a twenty thousand-person city marketed to that city, who knows what would develop for the farmers living a hundred miles from a Coke machine?

In my experience, those isolated rural farms and ranches have opportunities that I find difficult to take advantage of living near a small city. More isolated farms can get by with things like adding buildings or chuck wagon experiences (agri-tourism in which people come out for a horseback trail ride and gather around a chuck wagon for dinner) because all their activities aren't monitored by nearby nosy neighbors who tattle on them to local building, business, and food police. If you put an outhouse on a three thousand acre Idaho ranch it doesn't excite the kind of scrutiny it does on the outskirts of Baltimore.

The trick is to find what kind of resource-, geographical-, or personnel-advantage your farm has, and then exploit it. Our farm wouldn't be a good place to host quail hunting, for example, because we're too close to neighbors. Every place has its assets and liabilities.

If you want your farm income to increase, think people first. People have money. Most agribusiness thrives on farm wealth extraction; successional farms thrive on people. Get people to your farm. Sell directly to people. People are where the wallets are. Instead of complaining about people moving into your rural area, embrace them and pick their pockets by offering goods and services other farmers are either too stodgy or paranoid to offer. I find it much easier to help someone part with their money when I'm hugging them — it's my personal wealth redistribution plan.

A whole book could be written about direct marketing, and I do intend to write one some day, but for today, suffice it to say that direct marketing creates many income-producing opportunities. Most farmers never think about this because most farmers really don't like people — that's why they're farmers. They're more friendly with their tractor than their spouse and they carry on better conversations with their cows than other people, especially *city* people! But city people have the money, and reaching out to them directly can yield tremendous benefits.

A farm capitalizing on direct marketing can generate substantial income without the farm itself being particularly large. Increasing the margin per unit through direct marketing frees farmers from striving after the bigger-bigger-bigger-bigger treadmill. Tiny margins push growth; wide margins take the pressure off.

That brings us to the next principle: scalability. Ideally, these complementary enterprises scale down as well as up. How small can they be and still be viable? How many cut flowers does it take to make a salable bouquet? Rather than thinking about how big it can be, we need to think about how small it can be and still be viable. Young people are attracted to opportunities they can actually envision. If it has to be a monstrosity to generate a salary, it's intimidating and it's not likely to happen.

The smaller the prototype can be, the less risky the innovation. The smaller the embryo, the more likely it will be birthed. Reducing overhead is key to getting the new enterprise off the ground. Again, one of the beautiful things about pastured poultry is that it's viable on a tiny production scale. The cash flow is fast so the enterprise can finance its own expansion.

Do you have ponds on your farm? How about having a young person layer on a recreational fishing enterprise? It might require weed whacking around the pond and a floating aerator to improve water quality, but otherwise, you're just utilizing an underutilized resource. How about a solarium on the south side of the house? Extended season production captures the highest value on local produce because it's out of season. The house is already there. The sun is already there. You're just leveraging underutilized space and sunshine to create another income opportunity.

On our farm, we run beef cattle, eggmobiles, turkeys, and broilers across the same pasture acreage — at different times of the season, of course. But the stacking of those enterprises yields nearly ten thousand dollars per acre in gross annual sales. Compared to the nominal two hundred per acre for the typical beef-only farmer in our community, this increased income suddenly opens up a salaried opportunity to the next generation.

This brings us to our next principle, one I've emphasized again and again in this book: portable infrastructure. When we think about a farm, our minds are drawn to the infrastructure that a farmer brings to a piece of land. The difference between a farm and a wilderness area, for example, is not the terrain, rainfall, or climate; it's what the farmer has built or developed on the farm. A corral. A barn. A shed. A tractor.

In modern industrial agriculture, most of the farmer's infrastructure is stationary. Anybody see a portable Concentrated Animal Feeding Operation (CAFO) lately? I don't think so. But in a succession-friendly farm, the infrastructure is portable. That way enterprises can be shifted around from place to place. The shelters and control schemes (fences) don't dominate the landscape and don't break the bank.

Indeed, the equity in a succession-friendly farm is not in the land; it's in the management experience and the customer base. I can assure you that at Polyface the most valuable portion of our business is our customer base. The beautiful thing is that they're portable too — to a certain degree. They'll follow you from land base to land base as long as you don't stray too far. Portable infrastructure enables us to set up shop, so to speak, anywhere. We're not tied to a place; we're not limited to occupying a certain space.

Low-cost portable infrastructure doesn't enslave the next generation into doing the same things the previous one did. What do you do with a CAFO when the young person wants to change course? Imagine a confinement dairy farmer's son or daughter returning home from a grass-based dairying conference. The enthusiastic youngster presents a proposition to Grandpa: "Instead of planting all this corn and wheat, then mechanically harvesting it, storing it, feeding it and then hauling all the manure back out to the fields, why don't we just let perennial grass grow and move the cows around with electric fence, letting them self-harvest and self-fertilize?"

Of course, Grandpa responds incredulously, midst apoplectic seizures. "What?" he says. "I spent my whole life pouring concrete, building silos (bankruptcy tubes) and putting together the best machinery money could buy, and you . . . you. . . . you're just going to walk away from it?"

Pregnant pause: "Whose child are you?"

When farms invest heavily in single-use stationary capital infrastructure, it enslaves the farm and its descendants to continue using that infrastructure in the same way. The enslavement is both emotional and economic. These investments represent the sum and substance of the previous generation's life; they're monuments to personal accomplishment. It's hard to walk away from things like that.

When the next generation feels beholden — emotionally or monetarily — to the previous generation's paradigm, the future looks bleak. Young people need open-ended opportunity. They need to know that scrapping the old paradigm won't break the bank, or the brain. Few things encourage young people to flee faster than feeling locked into the same-old-same-old. Innovation is the next generation's lifeblood.

Most great innovations occur not as small tweaks to an existing model, but by throwing out the whole model. Going from a CAFO to a pasture-based outfit is a radical, hundred and eighty-degree shift. Going from annuals to perennials, commodity marketing to direct marketing, and from being a hermit to a host of colorful and vibrant agri-tourism are all opposite models. Freeing up the next generation to explore these paradigm shifts is the best way to excite young people to pursue farming. Being stuck doesn't engender excitement, build business, or carry on a family enterprise.

Sometimes after a discussion like this with a roomful of farmers, I feel a defeated spirit coming back at me. The vibe is that these notions are all too different and new and terrifying. Generally farmers more than forty-five years old (obviously I'm using an arbitrary age — it varies from person to person) resist these ideas if they've only known commodity, single-species, high capitalization farming. After all, Big Ag demonizes and pooh-poohs these ideas as not being real agriculture.

The implication is that you're not a real farmer if you don't have a CAFO or combine. One of my very first presentations, back in the early 1980s, was to a Ruritan club in our county. I showed our farm slides (on a carousel projector — remember those?) and went through the whole song and dance. One old codger in the front row sat with a scowl on his face, arms folded across his chest, through the whole performance. At the end, I asked for questions and he jumped right in:

"You don't plow?"

"No, sir."

"You don't fill silos?"

"No sir."

"You don't plant corn?"

"No sir."

"Well, then, sonny, you don't do any farmin', do you?"

Modern mainline agriculturalists have a paradigm, and it's certainly not the one I just painted. Their antipathy is reasonable since mainline agriculture thrives on corporate income extracted at the expense of farmers' profits. When people ask me why my ideas aren't more widely adopted, my stock answer is that it would completely invert the power, position, prestige, and profits of the entire food and farming sector. That's a big ship to turn around, a whole lot of inertia.

The demise of farming communities throughout middle America, what's commonly referred to as "fly over country," is largely due to integrative and diversified farming principles being abandoned wholesale. When families couldn't stay on the farm, they couldn't stay in their communities. As a result, hundreds of rural towns have dried up. Dilapidated and composting farm houses stand as stark testaments to an economic model that pulls people off the farm rather than pulling them onto it.

To carry out all these jobs at our Polyface Farm takes a lot of people. When you drive up the lane in the mid-afternoon you'll see lots of activity. We have people on the phone taking orders. People moving chickens. People spreading compost. People harvesting vegetables. People cooking for the crew. People leading school children on tours. It's an active place.

I appreciate the fear I feel from forty-five-year-old farmers. New is always scary. But so is aging without young people. So is being the last leaf on the tree. To punch through that fear, I offer this: You don't personally have to do it. Should I repeat that? *You* don't have to do it. None of what I've described here is your responsibility. That's the liberating truth. All you have to do is recognize these principles as an effective way to bring youth back to the farm. That's all.

You don't have to start direct marketing. You don't have to pull chicken shelters under your apple trees. You don't have to make cornbread muffins from your corn. All I'm asking — encouraging — is that you open your heart to let a young person do these things on your farm. This doesn't mean you vacate the premises. And it doesn't mean you're being pushed aside.

Rather, it means your farm can enjoy the miracle of new birth — the birthing of a new farmer. The whole point here is that farms today are obviously not fully utilized at all. How do you fill up a farm? Farms are like soft-sided luggage — they can always take one more pair of socks. Think of these various people and enterprises as additional socks. Do you have a woodlot? Don't just sell the trees to a logging company for a few thousand dollars. How about partnering with a wood crafter and a bandsaw mill to create two incomes in perpetuity? Maybe a young person would like to start a firewood business from the poor quality stuff currently rotting for lack of use. You can work out arrangements with each enterprising entrepreneur — some sort of royalty — and enjoy being the wise old chief.

Farms have tons of room for additional enterprises. To fill the farm requires human energy and ingenuity. Stacking additional complementary enterprises offers a way to create additional salaries and bring more young people to the farm. That's what they call a win-win, because in your golden age, relaxing and watching the sun set on the farm you've loved — even as the sun rises there for an active new generation — will stir your heart with overwhelming joy and soothe your spirit with the promise of the cycles and the seasons built into your very lifeblood as a farmer. This is how we create fields of farmers.

Summary

F ood and fiber needs coupled with strained resources
and aging farmers presents unprecedented farming
opportunities. From urban roof-top farms to expansive
mob-stocking livestock operations in the Dakotas, every
nook and cranny of our world screams for renewal and human
massage.

In addition to traditional agrarian pursuits, today's new
societal needs offer even more farming types. I recently visited a
farm that sells calming and affirming emotional support for autistic
adults in Ohio. In my lifetime, autism ratios moved from 1:1,000 to
1:90. That's an exponential increase, and certainly part of what is
straining mental health services nationwide.

Unlike sheltered workshops, where low IQ people stuff
widgets in boxes all day, a farm servicing autistic adults offers active
work that better fits the needs autistic adults have to move around.
The therapeutic and educational opportunities available on farms
cannot be duplicated with pharmaceuticals or institutional settings.

Staunton, our queen city of the Shenandoah Valley, located
just 10 miles from our farm, hosted a couple of sanitoriums as early
as 1900. Known informally as "insane asylums," coloquially as
the "crazy house," more often these carried the moniker "funny

farm." Why attach farm to them? Because prior to the advent of sophisticated mind-altering pharmaceuticals, these facilities grew their own food. Dairy cows and vegetables provided meaningful work and often contributed remarkably to patient stability.

Animals and plants don't yo-yo emotionally like some humans. Modern studies show the same therapeutic benefits among elderly folks who have pets. Tending plants and animals gives a person meaning, purpose, and need. Everyone craves to be needed. Some 40 years ago the doctors who ran these facilities determined that asking the patients to tend gardens and gather eggs was abusive.

New mental hospitals sprang up, freed from farming operations and shackled instead to regimens of mind-altering drugs. At that time I interviewed the administrator of one of these institutions, asking him about the demise of the farm, and he declared vehemently: "every single mental illness can be cured with drugs." He absolutely and flatly refused to recognize any merit to tending plants and livestock.

In more recent times, the hospital is being rebuilt with more open spaces, less institutional trappings, and pathways back into supervised community involvement. Rather than just locking these patients up, a renewed interest among mental health professionals involves incorporating these patients into meaningful life experiences. I'm not trying to turn this book into a commentary on mental health treatment, but I think it's important to realize how extensive the modern agrarian opportunity really is.

I asked the farm manager taking me on a tour of this farm for autistic adults how he saw the need developing in the coming years, and he said it was beyond comprehension. As these impaired adults find a semblance of functionality and personal development in these special therapeutic farms, their popularity will spread. Not only do the patients enjoy the ambiance of the workplace, the primary caregivers find their job easier and more rewarding. Dare I say that the plants and animals seem to step up to the task as well, rewarding these folks with unconditional affection and production?

The Wounded Warrior project now includes sustainable and pasture-based farming as a gateway to independence. Many of these Wounded Warriors suffer emotionally and mentally--their wounds are not always physical. When people don't understand or respond to these injuries with prejudice and insults, plants and animals never pass judgment and are always happy to see people. Plants and animals don't mind if their caretakers are burned, maimed, or mentally suffering.

A network called Teaching Farms now spans the nation, offering school children real life farm experiences. I had the privilege of visiting one of these on Martha's Vineyard, where the youngsters pulled pastured poultry shelters across the field and then processed the birds for a local chef who prepared a gourmet meal to consummate the chicken experience. These youngsters, aged 9-12, looked me in the eye, shook hands firmly, explained their projects with informed knowledge that only comes by getting your hands dirty.

For these farms, the real product is experiential education, but in a day of profound disconnection with visceral living, what Richard Louv has dubbed "nature deficit disorder," perhaps selling something less tangible is certainly as important as selling eggs or cabbage. Our culture screams for real life experiences. Why is the extreme vacation, from mountain climbing to solo multi-day backpacking gaining such momentum? I think it's because nobody participates in the visceral foundations of life.

This fundamental separation screams for reconnection, and I suggest that farms are the best nexus for this reconnection to our ecological umbilical. These varied farms, then, that sell therapy, experience, connection offer opportunities that a more balanced civilization of the past did not need. Goodness, in the not too distant past, if you wanted to go somewhere you saddled up a horse. Today, not only have most young people never touched a horse, they haven't even eaten raw carrot or picked a green bean in the garden.

In a day of fantasy everything--work, recreation, money-- the need to get dirt under our fingernails has never been greater. Perhaps it bears saying, as well, that our world's ecology is tired of fantasy. When I walk the ground, it is not asking for more focus groups, legislative initiatives, or bureaucratic paperwork. The land wants my touch. It needs my touch. Not to continue the deranged and insulting touches of past generations, but the healing touch of today's penitent practitioner.

Summary

Who will touch this land in the future? Will the perfect storm created by aged-out farmers and ignorant youth move us toward more dysfunction, with larger farms, consolidated school districts, ghost towns in rural America? Will the land be gobbled up by large agri-business entities? Will the land be turned into playgrounds for the very rich, the only ones who can afford to own it?

Or will this storm create a revival of integrated, soil building, land healing, water generating, carbon sequestering, diversified, collaborative, multi-generational farms? The old saying "there ain't no money in farmin'" is a tired judgment on a tired paradigm uttered by tired hermit curmudgeons.

It is time to repopulate our farms with intellectual Jeffersonian agrarians. The youthful enthusiasm and innovation expressed in a thousand different ways by today's young people proves their readiness to take on the food and farm healing desperately needed. To you young people, I assure you that farming can absolutely produce a white collar salary from a pleasant life in the country.

If such an avocation thrills you, realize that you must go after it. You must chase that dream. It will not fall into your lap, pre-packed with a pretty Christmas bow on top. Start something. Do something. Pursue something. If you know a farmer you'd like to work with, give him this book. Donate a few hours a week. Soften him up. He's a tough old self-reliant coot. Romance him; break him down; endear yourself to him with service and dependability. Don't be a victim; don't assume entitlement to anything.

Pioneers didn't ask for freebies from the earth. They roughed out lean-tos, ate the same thing day after day, and built what you enjoy today. We need that spirit expressed in a thousand different hamlets today. With youth's indomitable spirit, you can handle your failures and setbacks. You have time to recuperate.

To you middle-aged farmers, please consider what life will be like to be 70 and alone. No doubt you've seen it in your community already countless times. Perhaps you have several pieces of equipment or shop tools purchased at estate auctions nearby. I beg you to consider some of the principles in this book and dare to look at your farm as a gold mine of opportunity.

I know you're working hard. I know the government makes it harder. But start today seeking and grooming at least one young

305

person to stack a symbiotic enterprise on your land. Maybe it'll be beehives. Maybe a garden on the edge of your corn field. Maybe a flock of pastured chickens on the lawn around the confinement hog house. Goodness, even bamboo to soak up the nitrogen from the confinement hog house. Who knows what could be a great enterprise?

But please don't assume you've exhausted your possibilities. None of us is even close to approaching that wall. Look around. See a small orchard over there? See a cobb and clay bread oven over there? How about a pretty grape arbor/fish pond/flower garden combination for picturesque weddings? Do you enjoy draft power? How about training a pair of oxen to do events and birthday parties, pulling attendees around in a wagon?

Goodness, when you start brainstorming, you can't write fast enough. And if you're stuck, get the children, the spouse, and friends together and have a no-holds-barred brainstorming session. Imagine all the things people could possibly want to buy that could be done on your farm. And don't forget a haven for that mental health professional in your family--want 5 acres over there to do a therapy farm?

I predict that on our current trajectory of McDonald's food, pharmaceutical health care, electronic fantasy existence, and nature-less urbanization, the number of credible income streams pouring into farms will be beyond our wildest imagination. Anyone with a farm will be considered not just a hero, but a societal healer. Now, you middle-aged farmer, go turn a bright eyed bushy-tailed entrepreneurial young person loose as a partner on your place. You'll be glad.

Finally, to the toughest agrarian, the elderly farmer whose children have already flown the nest, including the widow or widower trying to hang on, wondering about the future, let me assure you that all is not lost. First, the elephant in the room concerns inheritance. Why give your farm to the children who will sell it quickly or flounder with it? Why give that much value to someone who has not expressed a talent in keeping or leveraging it?

This is not about love; it's about stewardship. Realize that inheriting the farm outright to someone who has no clue how to maintain it throws the land, which really needs a protector, into the lap of an enemy. These are strong words, but I think they create a framework to develop a land ethic. Sometime we need to realize that the land outlives people. God gave the Israelites the land of Canaan, what He called the "gift of good land." It outlived Abraham and Isaac. It outlived Joshua. What is it about the land that is special? It endures.

I think until we as farmers view ourselves as pilgrims passing through a land that endures, we cheapen this gift to simple real estate. We record deeds and put a valuation on it. But it's not dollars and cents. It's the stream where children built rock dams in the summer. It's the clutch of mallard eggs that hatched under the sumac bush on the edge of the pond. It's where we sweated as teens alongside Dad, beating the thunderstorm with the sweetest-smelling bales of hay we ever put in the mow. It's where our daughter picked her first bouquet of wildflowers during a check on the electric fence one evening, presenting them to her mother (your wife), with outstretched radiance.

This is not just business. It's not just money. It's life. Let me assure you that young people, young couples, by the thousands are ready to come alongside you. Maybe it's too late for your middle-aged kids. But others don't share their jaundiced views of farm life. What would you give to see a young family embrace your farm, love it, caress it, replace the broken-down corral fence that's beyond your ability these days? How would you like to see your fields filled with laughter again? To see a supple, strong, man or woman--or both-- tossing hay bales and digging potatoes?

I am a strong believer in the old Chinese proverb: "When the student is ready, the teacher will appear." I can assure you, seasoned farmer, the students are ready. Will you deny these young people their chance? Their dreams? Perhaps you're a bit pessimistic. Maybe things didn't work out like you thought they would 30 years ago. Certainly you would consider yourself a realist, not a pessimist. Fair enough.

But things have changed. Consider how our culture has changed. Consider how things have changed in your lifetime. Plastic water pipe. Electric fence. Portable saw mills. Facebook. Internet. Laptops. Drip irrigation. It's not the same world you started in. It's a wonderful, amazing, optimistic world. Will you give a young person a chance with this new stuff? See what he can do? Let her fly.

You may have time left to launch the most amazing farmer in the history of the world. Someone who will put your land into tomorrow's healing legacy. These young people attend sustainable agriculture symposiums. They have Peace Corps assignments around the world. They're often stuck in meaningless cubicles working for "the man" at the end of the expressway. Seek and you will find. Many of them are looking for you. Don't hide from them.

I believe we can actually re-populate our countryside and urban lots and surburban lawns with professional well-paid farmers. With portable infrastructure, direct marketing, food clusters, and management equity young farmers have never had a more friendly environment to start. Older farmers have never had more interest in their vocation. Let's get together and make tomorrow--after today's storm---dawn clean and sparkling.

Appendix: Historical Context

Note from Joel: When I began writing this book, I knew one of our apprentices, Noah Beyeler, enjoyed research. He reads voraciously and always has great perspectives on issues. Instead of bogging myself down in this research, I decided to ask Noah if he'd get me up to speed on a historical context. True to form, Noah went above and beyond the call of duty. Rather than excerpting it, I decided to put it in the book as a complete chapter, in its entirety, under his byline. What better example of symbiotic mentor-apprentice projects than this excellent chapter? Thank you, Noah.

Originating from the Latin *apprendere* (to apprehend, grasp) the term 'apprentice' hasn't strayed far from its original definition of apprehending or learning something. The term 'apprenticeship' is first mentioned in England in the 13th century, "appearing in an ordinance of the London Lorimers [a craftsmen making bits, spurs, or other small metal works] in 1261, fixing the term of service at seven years and forbidding one master to entice away another's apprentice from him."[1] While this is one of the earliest mentions, given that documents from those days are relatively few and far between, one

---

[1] O. Jocelyn Dunlop, *English Apprenticeship and Child Labour A History* (New York, The MacMillan Company, 1912), 29.

can assume that the practice dates to earlier than this. By the end of the 1200s, most of the guilds within London must have at least employed some system of apprenticeship, since by 1300 "there is an act of the Common Council of the City (London) which deals with the enrollment of apprentices."[2] If legislation regarding apprentices was already deemed necessary by 1300, we can probably assume that the practice dates to much earlier than just 1261, since it was by 1300 relatively standardized and normal in England.

This does not mean that this was the earliest this system of training was in place. There is scant (but some) information on organized training in crafts dating back to ancient Egypt, and Babylon, as well as trade organizations in the Roman Empire. It makes sense that some form of an apprenticeship has for millennia been the preferred way to pass down a craft or trade, but specific information is scarce and so our history starts well into the Middle Ages in Europe with the formation of craft guilds. Since the apprentice system in England revolved around being able to become a member of a trade guild, the history of guilds and that of apprentices go hand in hand.

Early guilds were a mix of trade union and religious association, and were formed in the likes of merchant guilds (groups of merchants holding select rights to doing business in certain areas and towns). Trade guilds arose out of these merchant guilds as a way for individual craftsmen to protect their common interests. Many guilds however began as religious or fraternal organizations, and pursued moral goals alongside economic or trade-based ones. So, seemingly a long-used method for training assistants, 'apprenticeships' became adopted, reworked, and to some extent standardized by craft guilds in the Middle Ages, and over the course of time was made more and more mandatory within guild-dominated trades. By the 1400s, a standardized system of apprenticeship had taken hold in many crafts within England, and in 1562 was made law under England's *Statute of Artificers*, and became the official method of training for England's industrial classes. This system then, began at the start of the 12th century, was fully formed in England by the 15 and 16th

---

[2] Dunlop, 30.

centuries, and had began to decline by the 18th century, reaching a state of non-existence by the mid 1800s.[3]

The Formation of the Apprentice System (1200-1450)

While today most people correlate 'child labor' with the ugly abuses of the industrial revolution and modern industry, children have had a place in the labor force long before the industrial revolution. Since there was only a small industrial working class in the middle ages, agricultural work occupied a large number of working children. However, documentation about agricultural laborers is relatively limited compared to that of industrial labor. In the Middle Ages the lower classes (those working in agriculture) were mainly un-free, working small farms often not owned by them. So while the following history is focused around those children apprenticed to the 'skilled' trades of the age, it is important to remember that this covers only a small percentage of working children, those involved in the 'industrial' (if one could call it that) population.

While today we often think of an apprenticeship as a way for someone to gain knowledge and understanding of a trade for their personal betterment, or for entry into an occupation, the full-scale formation of the apprentice system in the 13th and 14th centuries in England seems to have as much to do with the ability of guilds to use apprentices to protect their common interests, as it does with furthering the goals and interests of the apprentices themselves.

For a long time, apprenticing was not the only method for

---

[3] Others break up these periods differently. "When historians consider 'apprenticeship,' they often generalize in terms of three extended periods. These may broadly be characterized as that of 'guild apprenticeship', let us say from about the twelfth century to 1563, with the state underpinning much practice; the period of statutory apprenticeship, from 1563 to 1814 (with guilds slowly attenuating); and finally a great diversity of forms which might be summarized as 'voluntary' apprenticeship, often agreements between employers and unions, from 1814 to the present day." K.D.M. Snell, "The Apprenticeship System in British History: the fragmentation of a cultural institution" *History of Education* 25:4 (1996): 304.

entering a trade. Depending on the guild and the place, one could enter a trade through a number of venues, even just by proving to be a skilled enough tradesman. However, as guilds began to increase their control of trade within their various realms, they began to realize that through increasing their regulation of who could enter a guild, they could control and decrease competition within their craft. While "these associations were in their origin voluntary and it was quite possible for one or other group of craftsmen in some town to refrain from forming themselves into a guild,"[4] wherever guilds were formed, membership often became mandatory if one wanted to trade within that craft. Regardless of the reasons for this move to increase control over individual crafts and trade, increasing the regulation of apprentices was one means through which guilds enforced and controlled the exclusivity over their trade.

From very early on guilds began to supervise the instruction of the young workers who were employed by their members. By 1398, the Leathersellers' guild in London maintained that no one shall "work in the [craft] if they be not bound apprentice."[5] As guilds began to realize that the enforcement of a much narrower system of apprenticeship helped prevent competition, regulations such as these became more and more common, until by the end of the 15th century it was increasingly difficult for outsiders to gain entry into guilds except through apprenticeship or by being the son of a member.

While ensuring that apprenticing was the only method to gain entry into guilds, guilds furthered their monopoly on trade by limiting the number of apprentices each guild member could have at any one time. Depending on the trade, apprentices were limited to one, two or several apprentices at a time at a time. Through this regulation guilds not only could maintain protection from competition in their trade, but could also ensure a certain aptitude in each new era of craftsmen, thereby helping maintain a guild's reputation. The control of competition that this regulation of apprentices and

---

[4] Dunlop, 28.
[5] "Ordinance of Lethersellers", Article 7, in William Black, *History and Antiquities of the Worshipful Company of Leathersellers* (London, 1871).

guild membership allowed made for a slow but steady adoption of the apprenticeship/guild system throughout England. By the mid-15th century it was widely practiced. To practice a trade one almost certainly had been apprenticed in that trade.

While we think of the 'evils' of child labor as mainly those attended to the children involved, in the Middle Ages guild members recognized the evils excessive child labor could potentially cause to themselves. Excessive cheap labor could potentially ruin the profits of guild members themselves by lowering the cost of goods. The excessive regulation of apprentices by guilds, through increasing the number of years of training needed, limiting the number of apprentices per master, or making more restrictive rules and raising the fees apprentices had to pay, was undoubtedly a matter of "self protection of adult workmen" and there "can be little doubt that a spirit of monopoly was at work."[6]   That said, while this might have limited some from becoming apprentices, or closed many trades to the general population, in the undemocratic society of the times it did give a relatively small number of children a strong education in specific trades, and a decent start to life. These children, however, were often those who least needed a helping hand. "Apprentices were mainly the sons of families with the means to finance the investment in training. By and large, apprenticeship did not allow poor families to improve their economic status, but rather provided middling and upper class families economic opportunities."[7]

The Regulation of the Apprentice System (1450-1650)

The regulation of apprenticeships took a certain series of steps. First off were rules to maintain fair play, like forbidding members from stealing apprentices from each other. Then guilds began to regulate how apprentices were to be kept track of, to ensure the exclusion of un-apprenticed craftsmen. This enrollment (known as 'binding' or 'ordering') was kept in books maintained

---

[6] Dunlop, 46-47.

[7] Tim Leunig, Chris Minns and Patrick Wallis, "Networks in the Premodern Economy: The Market for London Apprenticeships, 1600 – 1749," *The Journal of Economic History* 71:2 (2009): 3.

by officers of the guild or of the town. Then came regulations on who could be apprenticed (based on age, background, etc.), which guild members could have apprentices (sometimes only those who were married) and the number of apprentices each member could keep. These regulations however, while often similar from place to place were still relatively local, with individual guilds deciding on regulations and enforcement (for example, apprentice terms were usually 7 years, but varied from a low of 5 to as many as 10 years). It was not until the mid-16th century that a national system was put into place.

The Statute of Artificers was a set of English laws passed between 1558 and 1653 that regulated labor throughout the country. It set wages for different classes of workers, restricted the movement of workers, and lay down rules and regulations for apprenticeships. The Act transformed what had been a voluntary institution into a national one, and took what had been dozens of local customs and regulations and formed a series of national rules (such as standardizing the seven year apprenticeship term).

While we've already looked at some of the reasons guilds sought to regulate apprentices, the newly forming English state had its own reasons. This Statute was seen as a way to promote national prosperity. The economics of the 1500s led people to believe that the best way to secure a market was not by underselling competition, but by producing better goods; hence a nationwide apprentice program. Not that we should assume the law didn't help out guilds: the wealthiest trades (merchants, goldsmithing) were now officially reserved for the sons of propertied men. And since the laws passed were essentially an extension of the regulations already set in place by the guilds themselves, they maintained the status quo already imposed by the guilds. The Statute of Artificers, however, did not keep everyone happy.

While not easy to get around Guild rules, and the now national law, illegal workmen existed. Throughout the 1500s there are cases of complaints by the guilds of craftsmen trying to practice outside the apprenticeship regulations. However, these complaints were not excessively common, and we can assume that the Guilds were strong enough, for the most part, to uphold their regulations and the *Statute*. This success was short-lived however.

The Decline of the Apprentice System (1650-1850)

Not long after it was implemented on a national level, the guild and apprentice system started to wane, partly due to resistance towards its constraints on individual freedom. As the number of people without qualification in trades increased, and since regulatory authority from a central government was lacking, guilds had a harder and harder time as increasing numbers of workers set up shop without having completed an apprenticeship. Complaints from the guilds grew apace. In 1649 "the Merchant Tailors of London complained that their trade was overrun with foreigners and free traders, who worked without any qualification, at lower wages," and from 1660 "onwards, the guilds slowly but perceptibly began to lose ground."[8] In fact, from here on, illegal workmen began to complain about guilds as much as guilds did about the workers, and open defiance of the Statue began to be common.

While perhaps people still saw the apprenticeship system as a good way to train workers, the monopolistic attitude of guilds helped make it unpopular. As Europe began a shift towards a favoring of capitalism and free trade, economic thought began to turn in opposition to the idea of 'closed' crafts and the monopolization of trade.

By the mid 18th century, non-observance of the act was common. Not only was "legal opinion in the seventeenth century gradually becoming adverse to the Statute of Artificers...based on the dogma that liberty of trade was a natural and common law right,"[9] but the increasing industrialization and growth meant entire trades could no longer just be centered in one town. Guilds began to lose their ability to retain control over members and workers. On top of this, with the advent of machinery (and as 'industry' began to replace 'craft') many employees and employers recognized that 7 years of training was no longer necessary, and the guild and apprentice system suffered a relatively complete breakdown by the end of the 18th century.

---

[8] Dunlop, 112.
[9] Dunlop, 121.

By the 1800s, the traditional apprentice system had all but disappeared for the reasons mentioned above, and pauper apprentices (a system closer to the 'child labor' we envision today) were sent to work in factories all around the country. And so as the guild and apprentice system dissolved on itself by this time, in 1814 the Statute of Artificers was repealed, legally allowing any one to practice any trade unhindered. This, however, does not mean we see the end of 'apprentices' in the industrial setting. But the term takes on a different meaning around the industrial revolution. In 1802 the *Health and Morals of Apprentices Act* (which included a 12 hour working day) was passed, an act that dealt more with the new pauper apprentice system. Before we look at the pauper apprentice system, however, what was life like for early apprentices?

Life for an Apprentice

Seven years was relatively standard practice for apprentice terms, although depending on the craft terms ranged from between 5-10 years (10 in the case of highly skilled trades such as goldsmithing). Because most of what can be gleaned about apprentices during the Middle Ages comes from the individual agreements written between Masters and potential apprentices, it's hard to put together a full account of what life was like for them. Conditions must have depended a lot on the trade children were apprenticed into, as well as the individual Master they were apprenticed to. Conditions must have varied tremendously between guilds and members.

That said, one interesting piece to note is the focus on the intimacy between Masters and apprentices that multiple authors take pains to note. While today most people think of an apprenticeship as a sort of on-the-job training program, in the past there was much more of a focus on the obligations between trainer and trainee. In "medieval times [there was] a strong bond between the master and youth who lived in his house as a member of his family...a close personal bond between the young people and their employers kept constantly before the notice of the latter their responsibility for the rising generation of workmen...and the formation of character and training for adult life and citizenship."[10]   While we have looked at the formation of the apprentice system as a way for guilds to protect themselves economically, on an individual level masters must have been very obligated to oversee not only the 'technical' training of apprentices, but their moral and cultural upbringing as well. This differs dramatically from the labor laws of the Industrial Revolution and the 1800's, which, while ensuring the responsibility of employers over the bodily wellbeing of their employees (through work hours, factory safety, etc.) did not "reconstruct that sense of responsibility for the young people's future" which one saw in the guild apprentice system.[11]

This doesn't mean that conditions for Apprentices in the 14th or 15th century were all that great. Working hours were likely from sunrise to sunset in winter, and longer than 12 hour days in the summer.[12] They received no wages or earnings, but were usually afforded food, housing, and clothing (all three likely of the plainest sort). But that probably more reflected the living conditions of the times than any mistreatment of apprentices. As practical members of the masters family, apprentices were probably treated the same as the other children in the house; while undoubtedly subservient to their masters, this was more because of their age and not their station in life and apprentices likely worked, ate, and chored alongside any of the sons in the house.

---

[10] Dunlop, 182.

[11] As Sarah Vickerstaff notes about apprenticeships in the mid 20th century, "a key feature of good apprenticeships in the post-war era was that the... complex interplay of individual motivation, family help, community backing and intergenerational support, as well as the obvious locational and labour market forces which made the opportunities available. Apprenticeships were most likely to be successful in the past when they were strongly embedded in the social relations and occupational structure of a local community." *'I was just the boy around the place': what made apprenticeship successful?* Journal of Vocational Education and Training 59:3 (2007): 342.

[12] The Statute of Artificers set the working day for labourers or artificers at 5 a.m. to 8 p.m. from the middle of March to the middle of September, and from sunrise to sunset the rest of the year. Dunlop, 175.

But again, conditions varied greatly between masters. During the height of this system, the Guilds had enough sway and authority to prevent the misuse of child labor; but this might have been just a byproduct of the Guilds' need to regulate apprentices and keep tabs on the numbers and whereabouts of those apprenticed. While these regulations served mainly the interests of the guild members, the fact that such oversight was necessary and common also helped the apprentices themselves. There is plenty of documentation of Guild members being fined for beating or mistreating their apprentices. Because Guilds were constantly checking in on apprentices and Guild members in order to uphold regulations, apprentices also had some semblance of recourse against poor Masters. Guilds had 'searchers' whose task it was to find bad work; in making these rounds, these searchers also necessarily were in a position to see that apprentices were being taken care of (or, at the least, not being neglected).

Apprentices started at various ages, although often during the young teenage years, and sometimes much earlier. In some instances, guilds limited the minimum age at which an apprentice could finish (for the London guilds pre-Statute of Artificers this was 24 years old). While not paid at the end of the terms apprentices were sometimes given a small sum of money, or clothes or tools to help him on his way once the terms were fulfilled. Apprentices were required to remain unmarried until the end of their apprenticeship, and often had certain qualifications in terms of class, age, physical attributes, etc. in order to even qualify for an apprenticeship. Apprentices had very little book-learning early on, and not until the 1700's did stipulations in the indenture terms begin to allow boys to attend writing schools on a more frequent basis. Once done with their seven-year term, apprentices could set up a shop for themselves (if they had the means) or, more likely, work for several years for wages as a Journeyman until they had the ability to buy tools and start their own workshop.

A Typical Apprentice Contract

"This Indenture made the sixteenth day of January in the Seaventh yeare of the reigne of our Sovraigne Lady Anne of Greate Brittaine France and Ireland Queene Defender of the faithex Anno quo Dom 1708 Betweene William Selman of the pish of Corsham in the County of Wiltes Husbandman and Richard Selman son of the sd William Selman of the one pte And Thomas Stokes holder of the pish of Corsham aforesaid Broadweaver of the other pte Witnesseth that the said Richard Selman of his owne voluntarie will and with the consent of said father William Selman Hath put himself an Apprntice unto the said Thomas Stokes and with him hath covenanted to dwell as his Apprntice from the day of the date hereof until the full end and terme of Seaven Yeares fully to be compleate and ended during which tyme the said Richard Selman shall well and faithfully serve him the said Thomas Stokes his master his secrets lawfully to be kept shall keep his Commandm lawfull and honest shall doe and execute hurt unto his said Master he shall not doe nor consent to be done Tavernes or Alehouses hee shall not haunt Dice Cardes or any other unlawfull games hee shall not use Fornication with any woman hee shall not commit during such tyme as he shall stay in his Masters service Matrymony with any woman hee shall not contract or espouse himself during the said Terme of Seaven years The goods of his said Masters inordinately hee shall not wast not to any man he lend without his Masters leave, but as a true and faithfull servant shall honestly behave himself toards his sd Master and all his both in word and deeds And the said Thomas Stokes doth for himselfe his Executors and Administration promise and Covenant to and with the sd William Selman and Richard Selman his Appntice to teach or cause the said Richard Selman to be taught and instructed in the trade Art science or occupacon of a Broadweaver after the best manner that he can or may wih moderate Correction finding and allowing unto his sd Servant meate drinke Apparrell Washing Lodging and all other things whatsoev fitting for an Apprntice of that trade during the said

term of Seaven yeares And to give unto his sd apprntice at the end of the sd term double Apparrell one suite for holy days and one for worken days…"[13]

Pauper Apprenticeship

While we have looked at apprenticeship as a method of training, as well as a method of consolidating power within the Guild structure, the concept of apprenticeship was also seen as a method of poor relief. From as early as the mid 16th century, Henry VIII passed laws requiring vagrant children to be arrested and bound to apprentices until adulthood (between 20-24). This binding out of children (whether those of 'vagrant' parents, bastard children, or those of parents simply 'overburdened') was sometimes known as the parish apprentice system, and was not overseen by the Guilds. It was not an apprenticeship with the primary goal of teaching children, but of removing them from poverty. While sometimes these children were taught trades depending on whom they were bound to, often they were engaged in agriculture/husbandry (or for women housewifery). In other words, while in theory the 'apprenticing' of poor children would give them a skill or trade that would help keep them from poverty, in reality this type of binding out often just was a sort of indentured servitude for those children sent in to it. In fact, while the two systems were entirely different, it was eventually the poor way in which pauper apprentices were treated, and the poor conditions in which they lived, that damaged the reputation of the apprentice system in general.

While parish apprentices existed for some time in England before the industrial revolution, they were not really an integral part of the production process. Whereas Guild apprentices were specifically trained in production, parish apprentices more often did menial labor for households or agricultural work. However, as industry more and more relied on parish apprentices to fill the ranks of factory labor, and as the Guild/apprentice system fell by the

---

[13] Dunlop, 353.

wayside, parish apprentices took on a real part of the production and manufacturing process.[14]

Apprenticeship in Early America

What did early apprenticeships in America look like? Because most of the early immigrants to America were from Europe, they inevitably brought European-based apprenticeships over with them. Most of apprentices in America during the colonial period, however, were through pauper apprenticeships; again, more similar to indentured servitude than to a craft apprenticeship. This seems mainly because of the lack of guilds in America. While undoubtedly there were apprentices in every trade, they probably agreed to terms on an individual basis with tradesmen, and were not overseen with the same degree of regularity and regulation that they were in England.

Similar to the pauper apprentice system in England, pauper apprenticeship in early America was used to help those children otherwise left at the margins by society. Bound to masters like guild apprentices, pauper apprentices likely had much less in the way of recourse in the case of poor treatment. Living and working in the household they were bound to, boys were usually released at age 21 and girls at age 16. Much more similar to the customs brought on by English Poor Laws of the 1600s than by the *Statute of Artificers* in the 1500s, these pauper apprenticeships lasted from the start of colonial settlement well into the 1800s in America.[15]

Since these times, apprenticeships have taken on a wide range of forms in countries throughout the world, sometimes under the behest of government programs, sometimes through and because

---

[14] For further reading see: Katrina Honeyman *Child Workers in England, 1780-1820: parish apprentices and the making of the early industrial labour force* (England: Ashgate Press, 2007).
[15] For further reading see: Ruth Herndon and Murray John, eds. *Children Bound to Labor: The pauper apprentice system in early America* (Ithaca: Cornell University Press, 2009).

of industrial needs, and sometimes on an individual basis. Some countries, such as Germany, still have very strong apprenticeship programs while others, such as ours, have seen apprenticeship systems come and go, often under the guise of 'vocational education' programs instituted in times of industrial need (as in during World War 1, or throughout the 1940's and 1950's). Regardless, apprenticeships have taken many forms, yet today we seem to associate them solely with learning a trade or skill. For better or worse, there was much more to the apprenticeships of 1513 than of 2013, and not just because it was a 7-10 year commitment then. We today seem to overlook the other obligations and learning opportunities implicit in such a familial and custodial mentor/mentee relationship. Perhaps looking to the apprenticeships of the past we can uncover some of the benefits and hazards of apprenticeships in the future.